DOT-VNTSC-NPS-10-05

EXTERIOR SOUND LEVEL MEASUREMENTS OF SNOWCOACHES AT YELLOWSTONE NATIONAL PARK

April 2010
Final Report

Prepared for:
National Park Service
PO Box 168
Yellowstone National Park, WY 82190

Prepared by:
U.S. Department of Transportation
Research and Innovative Technology Administration
John A. Volpe National Transportation Systems Center
Environmental Measurement and Modeling Division, RVT-41
Cambridge, MA 02142

Notice

This document is disseminated under the sponsorship of the Department of Transportation in the interest of information exchange. The United States Government assumes no liability for its contents or use thereof.

Notice

The United States Government does not endorse products or manufacturers. Trade or manufacturers' names appear herein solely because they are considered essential to the objective of this report.

# REPORT DOCUMENTATION PAGE			Form Approved OMB No.

Public reporting burden for this collection of information is estimated to average 1 hour per response, including the time for reviewing instructions, searching existing data sources, gathering and maintaining the data needed, and completing and reviewing the collection of information. Send comments regarding this burden estimate or any other aspect of this collection of information, including suggestions for reducing this burden, to Washington Headquarters Services, Directorate for Information Operations and Reports, 1215 Jefferson Davis Highway, Suite 1204, Arlington, VA 22202-4302, and to the Office of Management and Budget, Paperwork Reduction Project (0704-0188), Washington, DC 20503.

1. AGENCY USE ONLY (Leave blank)	2. REPORT DATE April 2010	3. REPORT TYPE AND DATES COVERED Final Report	
4. TITLE AND SUBTITLE Exterior Sound Level Measurements of Snowcoaches At Yellowstone National Park		5. FUNDING NUMBERS VX82 - HM355	
6. AUTHOR(S) Christopher J. Scarpone, Aaron L. Hastings, Gregg G. Fleming, Cynthia S. Y. Lee, Christopher J. Roof			
7. PERFORMING ORGANIZATION NAME(S) AND ADDRESS(ES) U.S. Department of Transportation Research and Innovative Technology Administration John A. Volpe National Transportation Systems Center Environmental Measurement and Modeling Division, RVT-41 Acoustics Facility Cambridge, MA 02142-1093		8. PERFORMING ORGANIZATION REPORT NUMBER DOT-VNTSC-NPS-10-05	
9. SPONSORING/MONITORING AGENCY NAME(S) AND ADDRESS(ES) John Sacklin PO Box 168, Yellowstone National Park WY 82190 307-344-2024 John_Sacklin@nps.gov		10. SPONSORING/MONITORING AGENCY REPORT NUMBER	
11. SUPPLEMENTARY NOTES Lead National Park Service Technical Resource: Shan Burson, Grand Teton National Park PO Drawer 170, Moose, WY 83012, 307-739-3584, Shan_Burson@nps.gov			
12a. DISTRIBUTION/AVAILABILITY STATEMENT		12b. DISTRIBUTION CODE	

13. ABSTRACT (Maximum 200 words)
Sounds associated with oversnow vehicles, such as snowmobiles and snowcoaches, are an important management concern at Yellowstone and Grand Teton National Parks. The John A. Volpe National Transportation Systems Center's Environmental Measurement and Modeling Division (Volpe Center) is supporting the National Park Service (NPS) with its on-going Winter Use Plan (WUP) program. As part of this support, acoustic measurements of twenty-five snowcoaches were made at three locations in Yellowstone National Park in January 2009. Measurement methodologies were guided by SAE J1161 with recommended improvements from 2008 measurements also performed for NPS by the Volpe Center. Data collected will be used to 1) refine and finalize snowcoach sound level testing procedures recommended from the 2008 Volpe Study, 2) determine which snowcoaches have the Best Available Technology (BAT) with respect to noise emissions, 3) determine any site-specific measurement bias, 4) develop a sound level versus speed relationship to determine if any measures could be taken for vehicles exceeding BAT requirements, and 5) develop OSV noise-distance relationships for use in a modified version of Federal Aviation Administration's (FAA) Integrated Noise Model (INM) developed for NPS in a previous study.

14. SUBJECT TERMS Noise Measurement, Parks, Snowmobiles, Snowcoaches, Snow, Sound Propagation, Ground Effects, Integrated Noise Model			15. NUMBER OF PAGES 198
			16. PRICE CODE
17. SECURITY CLASSIFICATION OF REPORT Unclassified	18. SECURITY CLASSIFICATION OF THIS PAGE Unclassified	19. SECURITY CLASSIFICATION OF ABSTRACT Unclassified	20. LIMITATION OF ABSTRACT

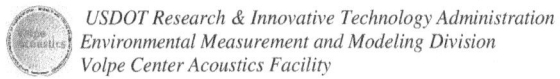

USDOT Research & Innovative Technology Administration
Environmental Measurement and Modeling Division
Volpe Center Acoustics Facility

April 2010

Table of Contents

Section — Page

Executive Summary 1

1 Introduction 1

2 Measurement Site Descriptions, Vehicles, and Equipment 3
   2.1   Measurement Sites 3
      2.1.1   South Entrance Site Location 4
      2.1.2   North Entrance Site Location 7
      2.1.3   West Entrance Site Location 10
   2.2   Measurement Conditions 13
   2.3   Vehicle Description 14
   2.4   Equipment Description 24
      2.4.1   Acoustic System 24
      2.4.2   Meteorological Equipment 25
      2.4.3   Vehicle Speed Collection 26

3 Measurement Protocol 27
   3.1   Methodology in Accordance with SAE International Standard J1161 27
   3.2   Measurements Supplemental to SAE J1161 Methodology 28
   3.3   Meteorological Measurements 30

4 Results and Analysis 31
   4.1   Meteorological Conditions 31
   4.2   Sound Level Results 31
      4.2.1   Ambient Sound Levels 32
      4.2.2   Sound Level Time History 32
      4.2.3   Maximum A-Weighted Sound Levels (L_{ASmx}) 35
      4.2.4   Sound Level Versus Speed 39
      4.2.5   Sound Exposure Level (SEL), dBA 41
      4.2.6   Sound Level Spectra 43
   4.3   Evaluation of Measurement Site Bias 45
      4.3.1   Comparison of 4 Foot Microphone Maximum Sound Levels 46
      4.3.2   Comparison of 4 Foot and 15 Foot Microphone Sound Levels at 50 Feet .. 49
      4.3.3   Barometric Pressure Effects on Measurements 49

5 Development of Data to Support Future Modeling 51
   5.1   Noise-Distance Curves 51
      5.1.1 L_{ASmx} Noise-Distance Curves 51
      5.1.2   SEL Noise-Distance Curves 54
   5.2   Spectral Data 56

6 Summary and Conclusions 59

Appendix A: Larson Davis Model 824 Sound Level Meter Settings 61

Appendix B: Sony Model TCD-100 DAT Recorder Settings 63

Appendix C: Measurement Protocol and Logging Procedure 65

Appendix D: Sound Level Time Histories 67

Appendix E: Overall Sound Levels 81

Appendix F: One-Third Octave Band Sound Levels 103

Appendix G: Measurement Site Bias Data 125

Appendix H: L_{ASmx} Noise-Distance Curves 135

Appendix I: Low Speed and High Speed SEL Noise-Distance Curves 147

Appendix J: Spectral Data for Modeling 159

References 173

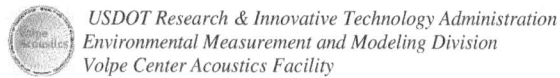

USDOT Research & Innovative Technology Administration
Environmental Measurement and Modeling Division
Volpe Center Acoustics Facility

April 2010

List of Figures

Section   Page

Figure 1. Measurement Site Locations Within Yellowstone National Park. 2
Figure 2. SAE J1161 Measurement Area Layout 4
Figure 3. Sound Level vs. Speed for All Snowcoaches 7
Figure 4. Measurement Site Locations in Yellowstone National Park. 4
Figure 5. Aerial Photo of the South Measurement Site (44.133, -110.66503). White dots indicate approximate location of microphones. 5
Figure 6. South Measurement Site Sketch – (A) Aerial View, (B) Profile (Not to Scale) 6
Figure 7. South Entrance Line-of-Sight from Microphone to Tracks 7
Figure 8. Aerial Photo of the North Measurement Site (44.881942, -110.732869). White dots indicate approximate location of microphones. 8
Figure 9. North Measurement Site Sketch – (A) Aerial View, (B) Profile (Not to Scale) 9
Figure 10. North Entrance Line-of-Sight from Microphone to Tracks 10
Figure 11. Aerial Photo of the West Measurement Site (44.6667, -110.9702). White dots indicate approximate location of microphones. 11
Figure 12. West Measurement Site Sketch – (A) Aerial View, (B) Profile (Not to Scale) 12
Figure 13. West Entrance Line-of-Sight from Microphone to Tracks 13
Figure 14. Purpose Built Bombardier B-12 with Tracks and Skis (Alpen Guide – Kitty) 15
Figure 15. Converted Ford E-350 Comm-Trans van with Mattracks (Buffalo Bus Touring #3) 15
Figure 16. Converted Ford F550 "Krystal" van with Grip Tracks (Buffalo Bus Touring – #4) . 16
Figure 17. Converted Ford E-350 Vanterra van with Mattracks (Buffalo Bus Touring – T2) 16
Figure 18. Converted Ford Econoline van with Mattracks 150 (See Yellowstone Tours – #4) .. 17
Figure 19. Converted Ford Vanterra with Mattracks 150 (See Yellowstone Tours – #6) 17
Figure 20. Converted Ford Odyssey van with Mattracks (See Yellowstone Tours – #9) 18
Figure 21: Converted Chevrolet Express Van with Mattracks (Xanterra – 430) 18
Figure 22. Purpose Built Bombardier B-12 with Tracks and Skis (Xanterra – 707) 19
Figure 23. Purpose Built Bombardier B-12 with Tracks and Skis (Xanterra 709*) 19
Figure 24. Purpose Built Bombardier B-12 with Tracks and Skis (Xanterra – 710) 20
Figure 25. Purpose Built Bombardier B-12 with Tracks and Skis (Xanterra – 713) 20
Figure 26. Purpose Built Pirnoth Powder Tour Cat TR with Rubber/Tracks (Xanterra Pirnoth – 537) 21
Figure 27. Converted Dodge B350 van with Snowbusters (Yellowstone Expeditions – Hayden) 21
Figure 28. Converted Ford E150 van with Snowbuster (Yellowstone Expeditions – Eleanor) .. 22
Figure 29. Converted Ford Econoline van with Mattracks (Yellowstone Snowcoach Tours – SNOWVAN4) 22
Figure 30. Converted Ford Econoline van with Mattracks (Yellowstone Snowcoach Tours – SNOWVAN5) 23
Figure 31. Acoustic System Setup 24
Figure 32. Instrumentation Case 25
Figure 33. TAMS Setup 26
Figure 34. SAE J1161 Measurement Area Layout 28
Figure 35. Extended Measurement Area Layout 29

Figure 36. Yellowstone Snowcoach (SNOVAN4) Van, Left Side at Low Speed (15 mph) 33
Figure 37. Yellowstone Snowcoach (SNOVAN4) Van, Right Side at High Speed (33 mph) 34
Figure 38. Yellowstone Snowcoach (SNOVAN4) Van, Left Side at Idle 34
Figure 39. Maximum Sound Level vs. Speed for All Snowcoaches 40
Figure 40. Xanterra 710, West Entrance (Jan 21st) Spectra for Low Speed and High Speed 44
Figure 41. Yellowstone Snowcoach – SNOVAN5, (Jan 14) Spectra for Low Speed and High Speed 45
Figure 42. Maximum Sound Level vs. Speed for Xanterra 713 47
Figure 43. Maximum Sound Level vs. Speed for Alpen Guide – Kitty 47
Figure 44. Maximum Sound Level vs. Speed for Yellowstone Expedition – Hayden 48
Figure 45. Maximum Sound Level vs. Speed for Yellowstone Snowcoach – SNOVAN5 48
Figure 46. Xanterra 713 Maximum Sound Pressure Level Noise-Distance Curves 53
Figure 47. Example Low Speed SEL Noise-Distance Curves 55
Figure 48. Xanterra 713, 1000 Feet, Maximum Spectra 57
Figure 49. Xanterra 709, South Entrance, (Jan 14th) Time History Plots 68
Figure 50. Alpen Guide – Kitty, South Entrance, (Jan 14th) Time History Plots 68
Figure 51. Yellowstone Snowcoach – SNOVAN5, South Entrance, (Jan 14th) Time History Plots 69
Figure 52. Yellowstone Expedition – Hayden, South Entrance, (Jan 14th) Time History Plots .. 69
Figure 53. Xanterra 713, South Entrance, (Jan 14th) Time History Plots 70
Figure 54. Alpen Guide – Kitty, North Entrance, (Jan 15th) Time History Plots 71
Figure 55. Yellowstone Snowcoach – SNOVAN4, North Entrance, (Jan 15th) Time History Plots 71
Figure 56. Xanterra 713, North Entrance, (Jan 15th) Time History Plots 72
Figure 57. Yellowstone Expedition – Hayden, North Entrance, (Jan 15th) Time History Plots .. 72
Figure 58. Xanterra 707, North Entrance, (Jan 16th) Time History Plots 73
Figure 59. Yellowstone Snowcoach – SNOVAN5, North Entrance, (Jan 16th) Time History Plots 73
Figure 60. Xanterra 537 (Pernoth), North Entrance, (Jan 16th) Time History Plots 74
Figure 61. Xanterra 430, North Entrance, (Jan 16th) Time History Plots 74
Figure 62. Yellowstone Expedition – Hayden, West Entrance, (Jan 20th) Time History Plots ... 75
Figure 63. Alpen Guide – Kitty, West Entrance, (Jan 20th) Time History Plots 75
Figure 64. Xanterra 713, West Entrance, (Jan 20th) Time History Plots 76
Figure 65. Yellowstone Snowcoach – SNOVAN5, West Entrance, (Jan 20th) Time History Plots 76
Figure 66. See Yellowstone Tours #6, West Entrance, (Jan 21st) Time History Plots 77
Figure 67. Buffalo Bus Touring #4, West Entrance, (Jan 21st) Time History Plots 77
Figure 68. Yellowstone Expedition – Eleanor, West Entrance, (Jan 21st) Time History Plots ... 78
Figure 69. Xanterra 710, West Entrance, (Jan 21st) Time History Plots 78
Figure 70. Buffalo Bus Touring T2, West Entrance, (Jan 21st) Time History Plots 79
Figure 71. See Yellowstone Tours #9, West Entrance, (Jan 22nd) Time History Plots 79
Figure 72. Buffalo Bus Touring #3, West Entrance, (Jan 22nd) Time History Plots 80
Figure 73. See Yellowstone Tours #4, West Entrance, (Jan 22nd) Time History Plots 80
Figure 74. Xanterra 713, South Entrance (Jan 14th) Maximum Spectra for Low Speed and High Speed 106

Figure 75. Yellowstone Expedition – Hayden, South Entrance (Jan 14th) Maximum Spectra for Low Speed and High Speed 106
Figure 76. Yellowstone Snowcoach – SNOVAN5, South Entrance (Jan 14th) Maximum Spectra for Low Speed and High Speed 107
Figure 77. Alpen Guide – Kitty, South Entrance (Jan 14th) Maximum Spectra for Low Speed and High Speed 107
Figure 78. Xanterra 709, South Entrance (Jan 14th) Maximum Spectra for Low Speed and High Speed 108
Figure 79. Alpen Guide – Kitty, North Entrance (Jan 15th) Maximum Spectra for Low Speed and High Speed 111
Figure 80. Yellowstone Snowcoach – SNOVAN4, North Entrance (Jan 15th) Maximum Spectra for Low Speed and High Speed 111
Figure 81. Xanterra 713, North Entrance (Jan 15th) Maximum Spectra for Low Speed and High Speed 112
Figure 82. Yellowstone Expedition – Hayden, North Entrance (Jan 15th) Maximum Spectra for Low Speed and High Speed 112
Figure 83. Xanterra 707, North Entrance (Jan 16th) Maximum Spectra for Low Speed and High Speed 113
Figure 84. Yellowstone Snowcoach – SNOVAN5, North Entrance (Jan 16th) Maximum Spectra for Low Speed and High Speed 113
Figure 85. Xanterra 537 (Pernoth), North Entrance (Jan 16th) Maximum Spectra for Low Speed and High Speed 114
Figure 86. Xanterra 430, North Entrance (Jan 16th) Maximum Spectra for Low Speed and High Speed 114
Figure 87. Yellowstone Expedition – Hayden, West Entrance (Jan 20th) Maximum Spectra for Low Speed and High Speed 118
Figure 88. Alpen Guide – Kitty, West Entrance (Jan 20th) Maximum Spectra for Low Speed and High Speed 118
Figure 89. Xanterra 713, West Entrance (Jan 20th) Maximum Spectra for Low Speed and High Speed 119
Figure 90. Yellowstone Snowcoach – SNOVAN5, West Entrance (Jan 20th) Maximum Spectra for Low Speed and High Speed 119
Figure 91. See Yellowstone Tours #6, West Entrance (Jan 21st) Maximum Spectra for Low Speed and High Speed 120
Figure 92. Buffalo Bus Touring #4, West Entrance (Jan 21st) Maximum Spectra for Low Speed and High Speed 120
Figure 93. Yellowstone Expedition – Eleanor, West Entrance (Jan 21st) Maximum Spectra for Low Speed and High Speed 121
Figure 94. Xanterra 710, West Entrance (Jan 21st) Maximum Spectra for Low Speed and High Speed 121
Figure 95. Buffalo Bus Touring T2, West Entrance (Jan 21st) Maximum Spectra for Low Speed and High Speed 122
Figure 96. See Yellowstone Tours #9, West Entrance (Jan 22nd) Maximum Spectra for Low Speed and High Speed 122
Figure 97. Buffalo Bus Touring #3, West Entrance (Jan 22nd) Maximum Spectra for Low Speed and High Speed 123

Figure 98. See Yellowstone Tours #4, West Entrance (Jan 22nd) Maximum Spectra for High Speed... 123
Figure 99. Alpen Guide – Kitty High Speed Time History Comparison.................................. 127
Figure 100. Alpen Guide – Kitty, Low Speed Time History Comparison 128
Figure 101. Alpen Guide – Kitty, Idle Time History Comparison 128
Figure 102. Yellowstone Snowcoach – SNOVAN5, High Speed Time History Comparison.. 129
Figure 103. Yellowstone Snowcoach – SNOVAN5, Low Speed Time History Comparison... 130
Figure 104. Yellowstone Snowcoach – SNOVAN5, Idle Time History Comparison............... 130
Figure 105. Yellowstone Expedition – Hayden, High Speed Time History Comparison 131
Figure 106. Yellowstone Expedition – Hayden, Low Speed Time History Comparison.......... 132
Figure 107. Yellowstone Expedition – Hayden, Idle Time History Comparison...................... 132
Figure 108. Xanterra 713, High Speed Time History Comparison 133
Figure 109. Xanterra 713, Low Speed Time History Comparison 134
Figure 110. Xanterra 713, Idle Time History Comparison 134
Figure 111. Xanterra L_{ASmx} Noise-Distance Curve 138
Figure 112. Yellowstone Expedition –Hayden L_{ASmx} Noise-Distance Curve 138
Figure 113. Yellowstone Snowcoach – SNOVAN5 L_{ASmx} Noise-Distance Curve 139
Figure 114. Xanterra 430 L_{ASmx} Noise-Distance Curve 139
Figure 115. Xanterra 537 L_{ASmx} Noise-Distance Curve 140
Figure 116. Xanterra 707 L_{ASmx} Noise-Distance Curve 140
Figure 117. Xanterra 707 L_{ASmx} Noise-Distance Curve 141
Figure 118. Xanterra 710 L_{ASmx} Noise-Distance Curve 141
Figure 119. Alpen Guide – Kitty L_{ASmx} Noise-Distance Curve.. 142
Figure 120. Yellowstone Snowcoach – SNOVAN4 L_{ASmx} Noise-Distance Curve 142
Figure 121. Yellowstone Expedition – Eleanor L_{ASmx} Noise-Distance Curve 143
Figure 122. See Yellowstone Tours #6 L_{ASmx} Noise-Distance Curve 143
Figure 123. See Yellowstone Tours #9 L_{ASmx} Noise-Distance Curve 144
Figure 124. Buffalo Bus Touring #3 L_{ASmx} Noise-Distance Curves... 144
Figure 125. Buffalo Bus Touring #4 L_{ASmx} Noise-Distance Curves... 145
Figure 126. Buffalo Bus Touring T2 L_{ASmx} Noise-Distance Curve.. 145
Figure 127. Xanterra 713 SEL Noise-Distance Curve... 150
Figure 128. Yellowstone Expedition – Hayden SEL Noise-Distance Curve 150
Figure 129. Yellowstone Snowcoach – SNOVAN5 SEL Noise-Distance Curve 151
Figure 130. Xanterra 430 SEL Noise-Distance Curve... 151
Figure 131. Xanterra 537 SEL Noise-Distance Curve... 152
Figure 132. Xanterra 707 SEL Noise-Distance Curve... 152
Figure 133. Xanterra 709 SEL Noise-Distance Curve... 153
Figure 134. Xanterra 710 SEL Noise-Distance Curve... 153
Figure 135. Alpen Guide – Kitty SEL Noise-Distance Curve... 154
Figure 136. Yellowstone Snowcoach– SNOVAN4 SEL Noise-Distance Curve 154
Figure 137. Yellowstone Expedition – Eleanor SEL Noise-Distance Curve 155
Figure 138. See Yellowstone Tours #6 SEL Noise-Distance Curve 155
Figure 139. See Yellowstone Tours #9 SEL Noise-Distance Curve 156
Figure 140. Buffalo Bus Touring #3 SEL Noise-Distance Curve 156
Figure 141. Buffalo Bus Touring #4 SEL Noise-Distance Curve 157
Figure 142. Buffalo Bus Touring T2 SEL Noise-Distance Curve... 157

Figure 143. Xanterra 713 Spectra at 1,000 Feet at the Time of L_{ASmx}.. 165
Figure 144. Yellowstone Expedition – Hayden Spectra at 1,000 Feet at the Time of L_{ASmx} 165
Figure 145. Yellowstone Snowcoach – SNOVAN5 Spectra at 1,000 Feet at the Time of L_{ASmx} 166
Figure 146. Xanterra 430 Spectra at 1,000 Feet at the Time of L_{ASmx}.. 166
Figure 147. Xanterra 537 Spectra at 1,000 Feet at the Time of L_{ASmx}.. 167
Figure 148. Xanterra 707 Spectra at 1,000 Feet at the Time of L_{ASmx}.. 167
Figure 149. Xanterra 709 Spectra at 1,000 Feet at the Time of L_{ASmx}.. 168
Figure 150. Xanterra 710 Spectra at 1,000 Feet at the Time of L_{ASmx}.. 168
Figure 151. Alpen Guide – Kitty Spectra at 1,000 Feet at the Time of L_{ASmx}........................... 169
Figure 152. Yellowstone Snowcoach – SNOVAN4 Spectra at 1,000 Feet at the Time of L_{ASmx} 169
Figure 153. Yellowstone Expedition – Eleanor Spectra at 1,000 Feet at the Time of L_{ASmx} 170
Figure 154. See Yellowstone Tours #6 Spectra at 1,000 Feet at the Time of L_{ASmx} 170
Figure 155. See Yellowstone Tours #9 Spectra at 1,000 Feet at the Time of L_{ASmx} 171
Figure 156. Buffalo Bus Touring #3 Spectra at 1,000 Feet at the Time of L_{ASmx} 171
Figure 157. Buffalo Bus Touring #4 Spectra at 1,000 Feet at the Time of L_{ASmx} 172
Figure 158. Buffalo Bus Touring T2 Spectra at 1,000 Feet at the Time of L_{ASmx} 172

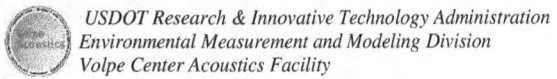

This page intentionally left blank

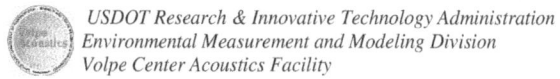

List of Tables

Section	Page
Table 1. Date and Location of Measurements in Yellowstone National Park 2009.	2
Table 2. Estimated Snow Cover by Entrance	3
Table 3. Snowcoach Vehicle Description	3
Table 4. Example Sound Level Calculation for SAE J1161	4
Table 5. L_{ASmx} for Loudest Side of Vehicle at Low Speed, dBA	5
Table 6. Date and Location of 2009 Measurements in Yellowstone National Park	3
Table 7. Estimated Snow Cover by Entrance	3
Table 8. Snowcoach Vehicle Description	14
Table 9. Instrumentation Operating Temperatures.	25
Table 10. TAMS Measurement Capabilities	26
Table 11. Example Sound Level Calculation for SAE J1161	28
Table 12. Summary of Meteorological Conditions During the Measurements	31
Table 13. Summary of Ambient Conditions During Measurements	32
Table 14. L_{ASmx} for Loudest Side of Vehicle at Low Speed, dBA	36
Table 15. L_{ASmx} for Loudest Side of Vehicle at High Speed, dBA	37
Table 16. Idle L_{Aeq} for Loudest Side of Vehicle at Idle (0 mph), dBA	38
Table 17. Average SEL for each Event at Low Speed, dBA	42
Table 18. Average SEL for each Event at High Speed, dBA	43
Table 19. Average Maximum Sound Level Difference Between 4 and 15 Foot Microphones ($L_{ASmx,4foot} - L_{ASmx,15foot}$) Low Speed and High Speed	49
Table 20. Summary of Temperature and Pressure for Measurement Periods	50
Table 21. Xanterra 713 Maximum Sound Level Noise-Distance Curve Values	52
Table 22. Xanterra 713 SEL Noise-Distance Curve Values	54
Table 23. Xanterra 713 Maximum Spectra at 1000 Feet	56
Table 24. Larson Davis 824 real time analyzer settings	61
Table 25. Low Speed Measurements used to Generate Final Reported L_{ASmx} Sound Levels for the South Entrance	82
Table 26. Low Speed Measurements used to Generate Final Reported L_{ASmx} Sound Levels for the North Entrance (part 1)	83
Table 27. Low Speed Measurements used to Generate Final Reported L_{ASmx} Sound Levels for the North Entrance (part 2)	84
Table 28. Low Speed Measurements used to Generate Final Reported L_{ASmx} Sound Levels for the West Entrance (part 1)	85
Table 29. Low Speed Measurements used to Generate Final Reported L_{ASmx} Sound Levels for the West Entrance (part 2)	86
Table 30. Low Speed Measurements used to Generate Final Reported L_{ASmx} Sound Levels for the West Entrance (part 3)	87
Table 31. High Speed Measurements used to Generate Final Reported L_{ASmx} Sound Levels for the South Entrance	88
Table 32. High Speed Measurements used to Generate Final Reported L_{ASmx} Sound Levels for the North Entrance (part 1)	89

Table 33. High Speed Measurements used to Generate Final Reported L_{ASmx} Sound Levels for the North Entrance (part 2) 90
Table 34. High Speed Measurements used to Generate Final Reported L_{ASmx} Sound Levels for the West Entrance (part 1) 91
Table 35. High Speed Measurements used to Generate Final Reported L_{ASmx} Sound Levels for the West Entrance (part 2) 92
Table 36. High Speed Measurements used to Generate Final Reported L_{ASmx} Sound Levels for the West Entrance (part 3) 93
Table 37. Idle Measurements to Generate Final Reported L_{Aeq} Sound Levels for the South Entrance 94
Table 38. Idle Measurements to Generate Final Reported L_{Aeq} Sound Levels for the North Entrance 94
Table 39. Idle Measurements to Generate Final Reported L_{Aeq} Sound Levels for the West Entrance 95
Table 40. Low Speed Measurements used to Generate Final Reported SEL Sound Levels for the South Entrance 96
Table 41. Low Speed Measurements used to Generate Final Reported SEL Sound Levels for the North Entrance 97
Table 42. Low Speed Measurements used to Generate Final Reported SEL Sound Levels for the West Entrance 98
Table 43. High Speed Measurements used to Generate Final Reported SEL Sound Levels for the South Entrance 99
Table 44. High Speed Measurements used to Generate Final Reported SEL Sound Levels for the North Entrance 100
Table 45. High Speed Measurements used to Generate Final Reported SEL Sound Levels for the West Entrance 101
Table 46. Maximum One-Third Octave Band Sound Levels for Select South Entrance Events (part 1), dB 104
Table 47. Maximum One-Third Octave Band Sound Levels for Select South Entrance Events (part 2), dB 105
Table 48. Maximum One-Third Octave Band Sound Levels for Select North Entrance Events (part 1), dB 109
Table 49. Maximum One-Third Octave Band Sound Levels for Select North Entrance Events (part 2), dB 110
Table 50. Maximum One-Third Octave Band Sound Levels for Select West Entrance Events (part 1), dB 115
Table 51. Maximum One-Third Octave Band Sound Levels for Select West Entrance Events (part 2), dB 116
Table 52. Maximum One-Third Octave Band Sound Levels for Select West Entrance Events (part 3), dB 117
Table 53. Low Speed L_{ASmx} for Snowcoaches at All Sites 125
Table 54. High Speed L_{ASmx} for Snowcoaches at All Sites 126
Table 55. Idle L_{Aeq} for Snowcoaches at All Sites 126
Table 56. L_{ASmx} Noise-Distance Curve Values (part 1) 135
Table 57. L_{ASmx} Noise-Distance Curve Values (part 2) 135
Table 58. L_{ASmx} Noise-Distance Curve Values (part 3) 136

Table 59. L_{ASmx} Noise-Distance Curve Values (part 4) 136
Table 60. L_{ASmx} Noise-Distance Curve Values (part 5) 137
Table 61. L_{ASmx} Noise-Distance Curve Values (part 6) 137
Table 62. SEL Noise-Distance Curve Values (part 1) 147
Table 63. SEL Noise-Distance Curve Values (part 2) 147
Table 64. SEL Noise-Distance Curve Values (part 3) 148
Table 65. SEL Noise-Distance Curve Values (part 4) 148
Table 66. SEL Noise-Distance Curve Values (part 5) 149
Table 67. SEL Noise-Distance Curve Values (part 6) 149
Table 68. Spectral Numerical Values for All Events at 1,000 Feet at the Time of L_{ASmx} (part 1) 159
Table 69. Spectral Numerical Values for All Events at 1,000 Feet at the Time of L_{ASmx} (part 2) 160
Table 70. Spectral Numerical Values for All Events at 1,000 Feet at the Time of L_{ASmx} (part 3) 161
Table 71. Spectral Numerical Values for All Events at 1,000 Feet at the Time of L_{ASmx} (part 4) 162
Table 72. Spectral Numerical Values for All Events at 1,000 Feet at the Time of L_{ASmx} (part 5) 163
Table 73. Spectral Numerical Values for All Events at 1,000 Feet at the Time of L_{ASmx} (part 6) 164

This page intentionally left blank

USDOT Research & Innovative Technology Administration
Environmental Measurement and Modeling Division
Volpe Center Acoustics Facility

April 2010

Executive Summary

Sounds associated with oversnow vehicles (OSV's), specifically snowmobiles and snowcoaches, are an important management concern at Yellowstone and Grand Teton National Parks. The John A. Volpe National Transportation Systems Center's Environmental Measurement and Modeling Division (Volpe Center) is assisting the National Park Service (NPS) with their Winter Use Plan (WUP) program (Ref. 1, 2, 3, 4) and supporting National Environmental Policy Act (NEPA) documents.

From January 14th though 22nd, 2009 the Volpe Center performed acoustic measurements of twenty five snowcoaches at three sites in Yellowstone National Park. Data collected will be used to:

1. Refine and finalize snowcoach sound level testing procedures recommended from a 2008 Volpe Study;
2. Determine which snowcoaches have the Best Available Technology (BAT), with respect to sound level;
3. Determine any site-specific measurement bias due to variables such as line of sight blockage, snow depth, barometric pressure associated with differing altitudes, etc.;
4. Develop a sound level versus speed relationship to determine if any speed restriction on snowcoaches is needed to meet BAT requirements; and
5. Develop OSV noise-distance relationships for use in a modified version of FAA's Integrated Noise Model (INM) developed for NPS in a 2006 Volpe Study[5]. In the 2006 study, the ground-to-ground propagation algorithms in INM Version 6.2[6,7] were modified to better account for sound propagation over snow-covered terrain. The modifications were based on the physical acoustics algorithms in the Federal Highway Administration's (FHWA) Traffic Noise Model (TNM)[8]. The process is described by Hastings et al. in Reference 9.

Sound level testing was conducted at three sites in Yellowstone National Park, one at each of the south, north, and west entrances. Table 1 shows the dates and locations of the measurements. Multiple sites were chosen to quantify any site biases that may exist. Site locations were provided by NPS and are shown in Figure 1. Table 2 shows the estimated snow cover for each entrance during the measurements.

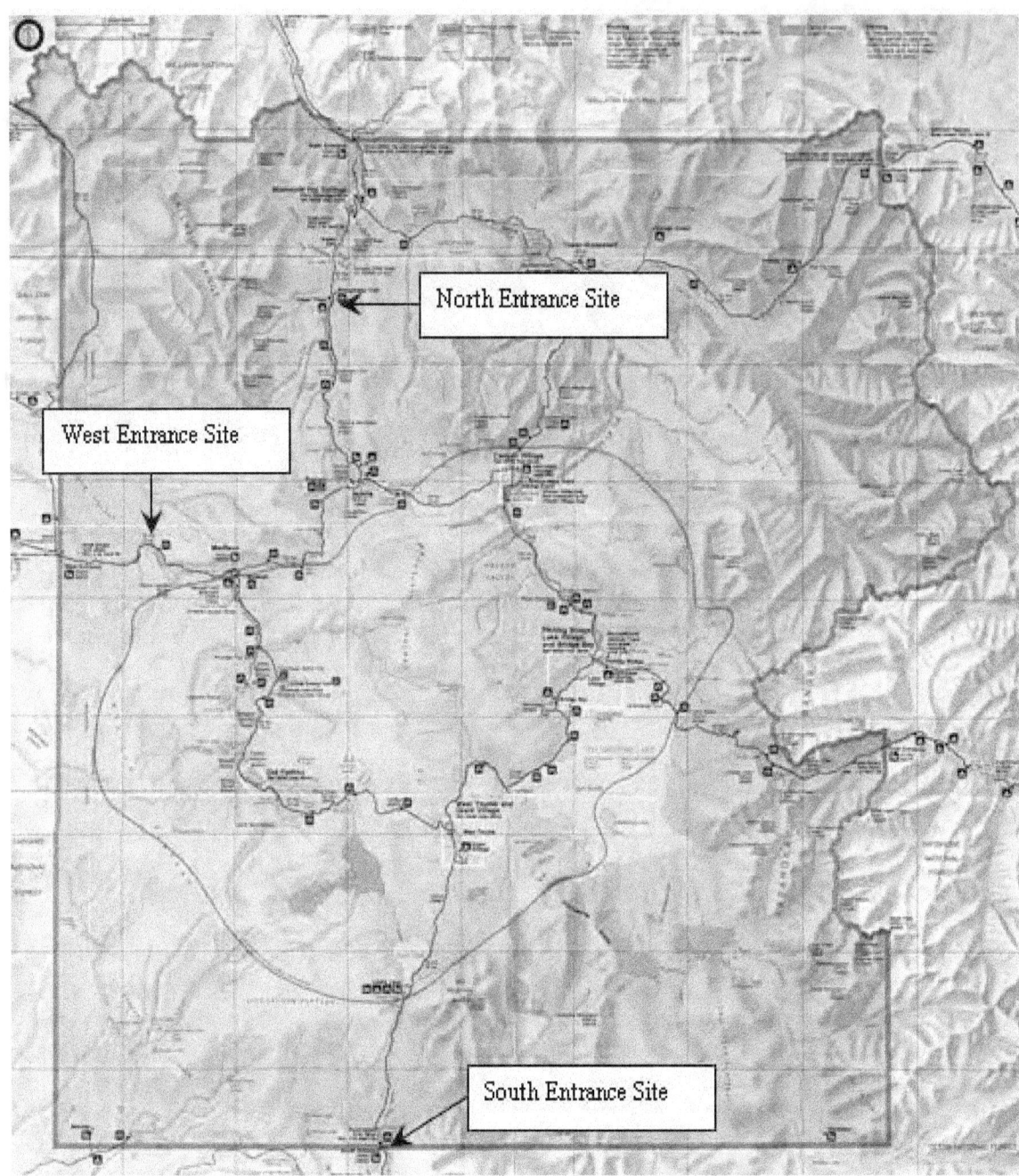

Figure 1. Measurement Site Locations within Yellowstone National Park.

Table 1. Date and Location of Measurements in Yellowstone National Park 2009.

Entrance	Date of Measurements
South	January 14th
North	January 15 - 16
West	January 20 - 22

Table 2. Estimated Snow Cover by Entrance

Entrance	Estimated Inches of Snow
South	30-42
North	12-24
West	12-24

Over the course of the six-day study, the team measured each side of 25 snowcoaches. The vehicles included 17 unique models; measurements were conducted at two constant pass-by speeds and idle. A summary of the snowcoach manufacturer and model type, track type, and engine characteristics is provided in Table 3. The Alpen Guide (Kitty), Xanterra 713, Yellowstone Snowcoach Tours (SNOVAN5), and Yellowstone Expedition (Hayden) were tested at all three sites.

Table 3. Snowcoach Vehicle Description

Snowcoach	Designation	Manufacturer	Model	Track Type*	Engine Year	Engine Size	Fuel Type	Passenger Capacity
Alpen Guide**	Kitty	Bombardier	B-12	Tracks & Skis	2002	5.3 L V-8 Chevy Vortek fuel-injected	Gasoline	10
Buffalo Bus Touring	#3	Ford	E-350 Comm-Trans	Mattracks	2006	6..9 L V-10	Gasoline	14
Buffalo Bus Touring***	#4	Ford	F550 "Krystal"	Grip Tracks	2009	6.4 L 8 Cylinder	Diesel	32
Buffalo Bus Touring	T2	Ford	E-350 "Vanterra"	Mattracks	2005	6..9 L V-10	Gasoline	14
See Yellowstone Tours	#4	Ford	Econoline	Mattracks 150	2000	6.8 L V-10	Gasoline	9
See Yellowstone Tours	#6	Ford	Vanterra	Mattracks 150	2004	6.8L V-10	Gasoline	13
See Yellowstone Tours	#9	Ford	Odyssey	Tank Tracks	2007	6.0 L	Diesel	20
Xanterra	430	Chevrolet	Express Van	Mattracks	2008	6.0 L V-8 GMC	Gasoline	12
Xanterra**	707	Bombardier	B-12	Tracks & Skis	2006	5.7 L V-8 GMC	Gasoline	10
Xanterra	709	Bombardier	B-12	Tracks & Skis	2006	5.7 L V-8 GMC	Gasoline	10
Xanterra**	710	Bombardier	B-12	Tracks & Skis	2006	5.7 L V-8 GMC	Gasoline	10
Xanterra**	713	Bombardier	B-12	Tracks & Skis	2006	5.7 L V-8 GMC	Gasoline	10
Xanterra	537	Pirnoth	Powder Tour Cat TR	Rubber/Pirnoth	2008	5.2 L V-8 Dodge	Gasoline	15
Yellowstone Expedition	Hayden	Dodge	B350 Van	Snowbuster wide track	1994	5.2 L V-8	Gasoline	10
Yellowstone Expedition	Eleanor	Ford	E150 Van	Snowbuster – Bombardier tracks	1997	4.6 L V-8	Gasoline	8
Yellowstone Snowcoach	SNOVAN4	Ford	Econoline	Mattracks	2000	6.8 L V-10	Gasoline	10
Yellowstone Snowcoach	SNOVAN5	Ford	Econoline	Mattracks	2001	6.8 L V-10	Gasoline	10

* Track types for the snowcoaches are as follows: Skis / Tracks – the rear wheels have been replaced with tracks while the front wheels have been replaced with skis; Mattracks – both the front and rear wheels have been replaced with triangular tracks manufactured by Mattracks; Snowbuster – are a specific manufacturer of Skis / Tracks, and Grip Trax – another specific type of triangular tracks to replace the front and rear wheels.
** These vehicles have been retrofitted for a reduced sound level.
*** Vehicle includes an air emission reduction device which operates periodically and creates additional noise. The device did not operate during testing, however at other times the pollution-control device was clearly audible. It is recommended that testing endeavor to include these types of sound sources.

SAE J1161 specifies a methodology for the measurement of exterior operational sound levels for snowmobiles. This methodology has been used in this study as snowmobiles and OSV's have similar propulsion systems, namely tracks and skis and are used at similar speeds in the parks. The methodology requires the measurement area to be an open region of packed snow at least 2 inches deep and free of reflecting surfaces. The snowpack is required to be sufficient to support the OSV without penetration and with no more than 3 inches of loose snow on top of the packed snow. An illustration of measurement area requirements is provided in Figure 2.

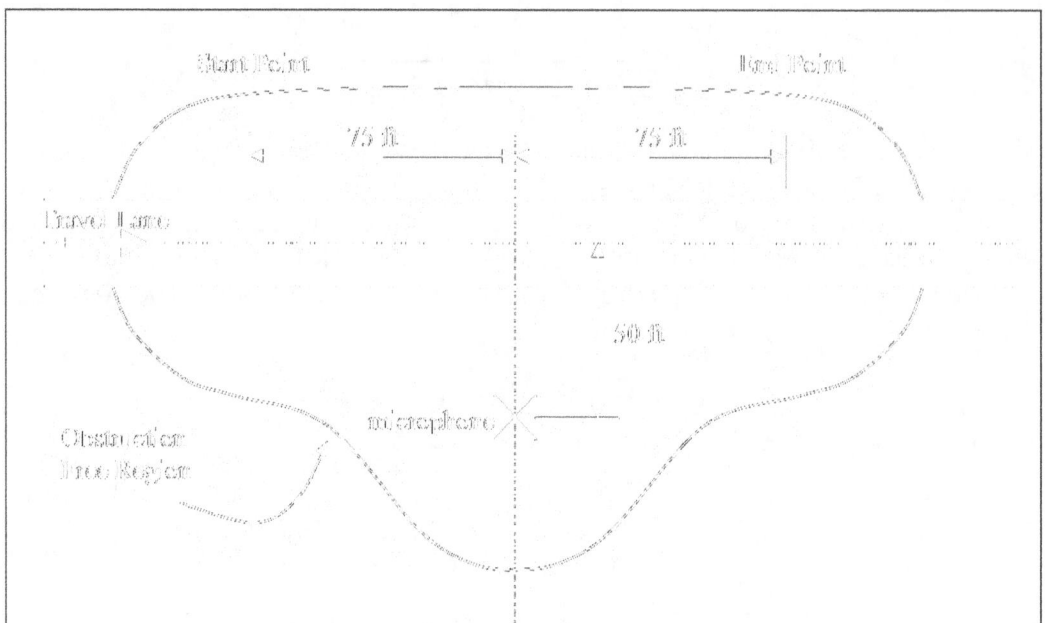

Figure 2. SAE J1161 Measurement Area Layout

SAE J1161 requires the OSV to approach the measurement area at a speed of 15 mph and maintain this speed throughout the area. The microphone and sound level meter are required to be positioned 50 feet from the center of the travel lane and 4 feet above the snow cover (see Figure 2). The sound level meter shall measure the maximum A-weighted sound pressure level with slow time weighting (L_{ASmx}) during each vehicle pass-by at a constant speed of 15 mph. Measurements are repeated until three L_{ASmx} values are measured within 2 dB. These values are then arithmetically averaged and rounded to the nearest integer. The measurements are conducted for the vehicle traveling in both directions; the side of the vehicle with the highest average is reported as the sound level for the OSV. The sides of the vehicle are designated as the "right side" and "left side" of the vehicle, from the perspective of the vehicles driver. An example of the sound level calculation is provided in Table 4.

Table 4. Example Sound Level Calculation for SAE J1161

Pass-by	L_{ASmx} Sound Level (dBA)			Within 2 dB?	Average	Final Sound Level
Left Side	64	65	65	yes	65	65
Right Side	62	61	62	yes	62	

Table 5 rank orders the twenty five snowcoaches from highest to lowest maximum A-weighted sound pressure levels measured with slow time weighting (L_{ASmx}) for constant low speed pass-bys [15 mph (nominal)] for the 50 foot microphone. The values listed in Table 5 are the arithmetically averaged L_{ASmx} values for the pass-by direction which yielded the highest average value. Averages of the measured speeds are specified in the last column.

Table 5. L_{ASmx} for Loudest Side of Vehicle at Low Speed, dBA

Vehicle	Entrance	Vehicle Side*	50 Foot Average L_{ASmx} (dBA)	100 Foot Average L_{ASmx} (dBA)	Average Speed of Runs (mph)
See Yellowstone Tours #9	West	Right	76	68	15
73 DBA BAT Limit					
Xanterra 537	North	Right**	71	65	10.7***
See Yellowstone Tours #6	West	Right	69	63	16
Xanterra 709	South	Left**	68	62	15
Xanterra 707*****	North	Right	68	62	15
Buffalo Bus Touring #3	West	Right**	67	59	15
Buffalo Bus Touring T2	West	Left	67	59	16
Xanterra 713*****	South	Right	66	60	16
Yellowstone Snowcoach – SNOVAN5	West	Left	66	59	15
Xanterra 713*****	West	Right	66	60	15
Yellowstone Snowcoach – SNOVAN5	North	Left	66	60	16
Buffalo Bus Touring #4	West	Left	65	59	16
Xanterra 430	North	Left	65	57	17
Yellowstone Expedition – Eleanor	West	Left	65	59	15
Xanterra 713*****	North	Right	65	59	15
Xanterra 710*****	West	Left	64	58	15
Yellowstone Expedition – Hayden	West	Right	64	57	16
Yellowstone Snowcoach – SNOVAN4	North	Left	63	57	15
Alpen Guide – Kitty*****	West	Left	61	55	16
Yellowstone Expedition – Hayden	North	Left	60	55	15
Yellowstone Snowcoach – SNOVAN5	South	Right	59	48	15
Alpen Guide – Kitty*****	North	Left**	59	53	15
Yellowstone Expedition – Hayden	South	Left	59	50	15
Alpen Guide – Kitty*****	South	Right**	58	47	15
See Yellowstone Tours #4	West	N/A****	N/A****	N/A****	N/A****

* "Left/Right" indicates left/right side of vehicle from the driver's perspective.
** Indicates that data was only available for one side of the vehicle.
*** The Xanterra 537 has a top speed of approximately 16 mph. This speed was captured for the development of a noise-distance relationship (see Section 5.1).
**** See Yellowstone Tours #4 did not have a low speed series due to the accumulation of snow on the windscreens.
***** These vehicles have been retrofitted for a reduced sound level.

The pass-by sound levels for low and high speeds were combined with idle sound levels to generate noise versus speed relationships as shown in Figure 3. It should be noted that evaluating the curves, based on idle, low and high speeds, is an imperfect method. Measuring more points would create a more realistic representation of the snowcoaches noise characteristics, but is cost prohibitive and impractical. Previous vehicle studies[10] have shown that an increase in speed versus sound level should roughly follow an increasing monotonic curve. This relationship would allow us to qualitatively describe gross site differences.

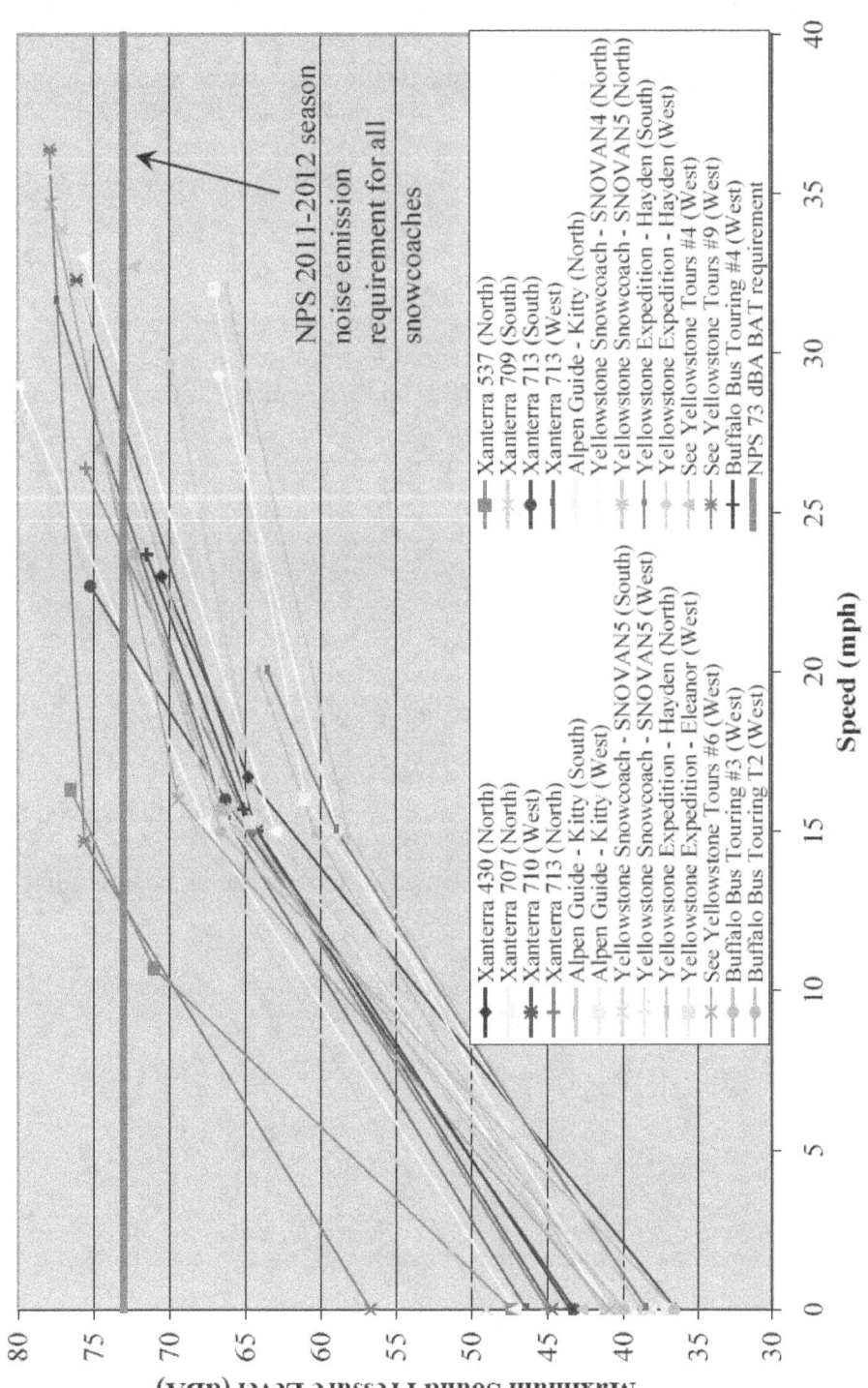

Figure 3. Sound Level vs. Speed[1] for All Snowcoaches

---

[1] Theoretically the sound level curves should asymptotically approach a constant sound level as the speed approaches 0 mph. However, measurements at more speeds would be required in order to capture the shape of this curve. For modeling purposes in the parks, three points are considered sufficient.

Based on the 2008 and the 2009 OSV testing at Yellowstone and Grand Teton National Parks, future snowcoach measurements in parks should adhere to SAE J1161 with the following modifications and considerations:

- Due to the altitude of the park, allowable barometric pressures during measurements should be expanded to include pressures typical of those experienced in the Yellowstone during the winter season[*]. The affects on sound levels due to barometric pressure could be accounted for in a manner similar to the methods described in References 11 and 12.
- Testing should be conducted for three conditions:
    - Idle
    - 15 mph
    - A high speed, to be determined by the park based on local speed limits, e.g., 30 mph, speed limit, or a typical cruising speed.
- Ambient measurements should be taken regularly (at least hourly) to accurately represent the changing conditions throughout the day.
- In order to standardize measurement sites the following criteria need to be followed:
    - Measurements should not be performed at sites with a snow berm. The 50 foot microphone should have a clear line of site to any potential noise source on the test snowcoach.
    - A groomer should be utilized to minimize track condition deterioration during testing. A limited number of coaches per day should be tested at any given location depending on the track conditions.
    - Site should be away from noise sources that could potentially interfere with data collection (e.g. running water),
    - Sites should be free of obstructions (e.g. trees and structures),
- If a vehicle fails to meet BAT requirements at the high speed, consideration should be given to:
    - Restrictions that would still allow the snowcoach to operate in the parks, but at reduced speeds;
    - Modifying the vehicles to reduce sound level; or
    - Removing the vehicle from the fleet and replacing it with a quieter one.

---

[*] This level could be set at 23.4 in. Hg (792 millibars) which is the standard in the BAT requirement for snowmobiles.

USDOT Research & Innovative Technology Administration
Environmental Measurement and Modeling Division
Volpe Center Acoustics Facility

April 2010

1 Introduction

Sounds associated with oversnow vehicles (OSV's), specifically snowmobiles and snowcoaches, are an important management concern at Yellowstone and Grand Teton National Parks. The John A. Volpe National Transportation Systems Center's Environmental Measurement and Modeling Division (Volpe Center) is assisting the National Park Service (NPS) with their Winter Use Plan (WUP) program (Ref 1, 2, 3, 4) and supporting National Environmental Policy Act (NEPA) documents.

The use of Best Available Technology (BAT) OSV's is one management approach to reducing the sound levels in the parks due to OSV's. An important NPS objective is to establish procedures to evaluate snowcoach compliance with BAT requirements[*]. Winter use regulations (Federal Register, Vol. 72, No. 239, December 13, 2007, pages 70781-70804) state that "Beginning in the 2011-2012 season, all snowcoaches must meet a sound emission requirement of no greater than 73 dBA. In order to determine which snowcoaches are BAT compliant the Superintendent will establish procedures for determining compliance" (36 CFR 7.131(4)(v)). Although these regulations were vacated by order of the U.S. District Court for the District of Columbia in September 2008, the NPS expects that snowcoach BAT will be a component of future winter planning.

In February 2008, the Volpe Center performed acoustic measurements of ten snowcoaches and six snowmobiles[13]. These measurements were made with two primary objectives: (1) help determine what testing protocols should be used to determine if *snowcoaches* meet the sound level BAT and; (2) determine which *snowcoaches* currently operating in the parks meet the sound level BAT standards. The Volpe Study resulted in a number of recommended improvements to the procedures in SAE J1161[14]:

- Because of the high elevation of the park, the following should be noted regarding barometric pressure specifications:
    - Measurements in the park will take place regardless of the barometric pressure, but the barometric pressure will be noted.
    - Future modeling results can be adjusted for barometric pressure.
- If a snow berm is present, all practical efforts to remove it should be implemented.
- If a snow berm greater than 3 feet in height cannot be removed, another site should be sought.
- Testing should be conducted for three conditions
    - Idle
    - 15 mph
    - A high/cruise speed, which may be limited by safety and/or posted speed limits.
- A groomer should be kept on hand in order to ensure that the track conditions do not deteriorate over the course of the testing.

---

[*] BAT classification also includes consideration of reduced and/or cleaner exhaust emissions.

- If vehicles fail to meet BAT requirements at higher speeds, consideration should be given to restrictions which would still allow the snowcoach to operate in the parks, but at a reduced speed.

From January 14th though 22nd, 2009 the Volpe Center performed additional acoustic measurements of twenty five snowcoaches at three sites in Yellowstone National Park. Data collected will be used to:

1. Refine and finalize snowcoach sound level testing procedures recommended from the 2008 Volpe Study[5];
2. Determine which snowcoaches meet the BAT with respect to sound level;
3. Determine any site-specific measurement bias due to variables such as line of sight blockage, snow depth, barometric pressure associated with differing altitudes, etc.;
4. Develop a sound level versus speed relationship to determine if any speed restriction on snowcoaches is needed to meet BAT requirements; and
5. Develop OSV noise-distance relationships for use in the modified version of FAA's Integrated Noise Model (INM) developed for NPS in 2006 Volpe Study[5]. In the 2006 study, the ground-to-ground propagation algorithms in INM Version 6.2[6,7] were modified to better account for sound propagation over snow-covered terrain. The modifications were based on the physical acoustics algorithms in the Federal Highway Administration's (FHWA) Traffic Noise Model (TNM)[8]. A journal article, Development of a tool for modeling snowmobile and snowcoach noise in Yellowstone and Grand Teton National Park[9], describes this process in detail.

Section 2 provides a description of measurement sites, and vehicles measured, along details of the measurement setup for the 2009 measurement study. Section 3 describes the measurement methodology, including details from the SAE J1161 standard. Section 4 presents data analysis and discussion. Section 5 presents data developed for use in noise modeling. A summary with conclusions and proposed next steps is included in Section 6. Appendices A and B contain details on the acoustic measurement equipment. Appendix C contains a step-by-step measurement protocol. Appendix D contains sound level time histories. Appendix E contains the individual event data used to obtain the overall snowcoach sound level. Appendix F presents the spectra that are associated with the maximum sound level results. Appendix G presents data comparing four specific vehicles tested at all the measurement sites. Appendices H and I present the low speed and high speed noise-distance curves, for modeling, for the maximum sound level and SEL, respectively. Appendix J presents the spectra measured for the low speed and high speed pass-bys. All related references are presented at the end of this document.

2   Measurement Site Descriptions, Vehicles, and Equipment

This section presents information on the layout and conditions of the measurement sites, the OSVs measured, and the measurement equipment used.

2.1   Measurement Sites

In January 2009 measurements were conducted in Yellowstone National Park, which is located in Wyoming, Montana, and Idaho. Sound level testing was conducted at three sites in Yellowstone National Park, one at each of the south, north, and west entrances. Table 6 shows the dates and locations of the measurements. Multiple sites were chosen to quantify any site biases that may exist. Site locations were provided by NPS and are shown in Figure 4. Table 7 shows the estimated snow cover for each entrance during the measurements.

Table 6. Date and Location of 2009 Measurements in Yellowstone National Park

Entrance	Date of Measurements
South	January 14th
North	January 15 - 16
West	January 20 - 22

Table 7. Estimated Snow Cover by Entrance

Entrance	Estimated Inches of Snow
South	30-42
North	12-24
West	12-24

Three sites were chosen with the intent of establishing future test locations near three entrance locations where many snowcoaches originate. NPS may potentially use these sites for future snowcoach testing. If snowcoach sound level BAT requirements are established, the locations would be convenient for nearby operators and pre-established for the NPS. In previous oversnow vehicle sound level testing, only the South Entrance location had been used.

At each of the three sites, two microphones were located 50 feet from the center of the travel lane, one each at 4 and 15 feet above the snow cover. An additional microphone was located 100 feet from the center of the travel lane, 4 feet above the snow cover. A meteorological system was set up 75 feet from the center of the travel lane in line with the microphones, 4 feet above the snow cover. An observer station was positioned approximately 100 feet off to the side of the microphone line and all data recording equipment was set up at this location.

USDOT Research & Innovative Technology Administration
Environmental Measurement and Modeling Division
Volpe Center Acoustics Facility

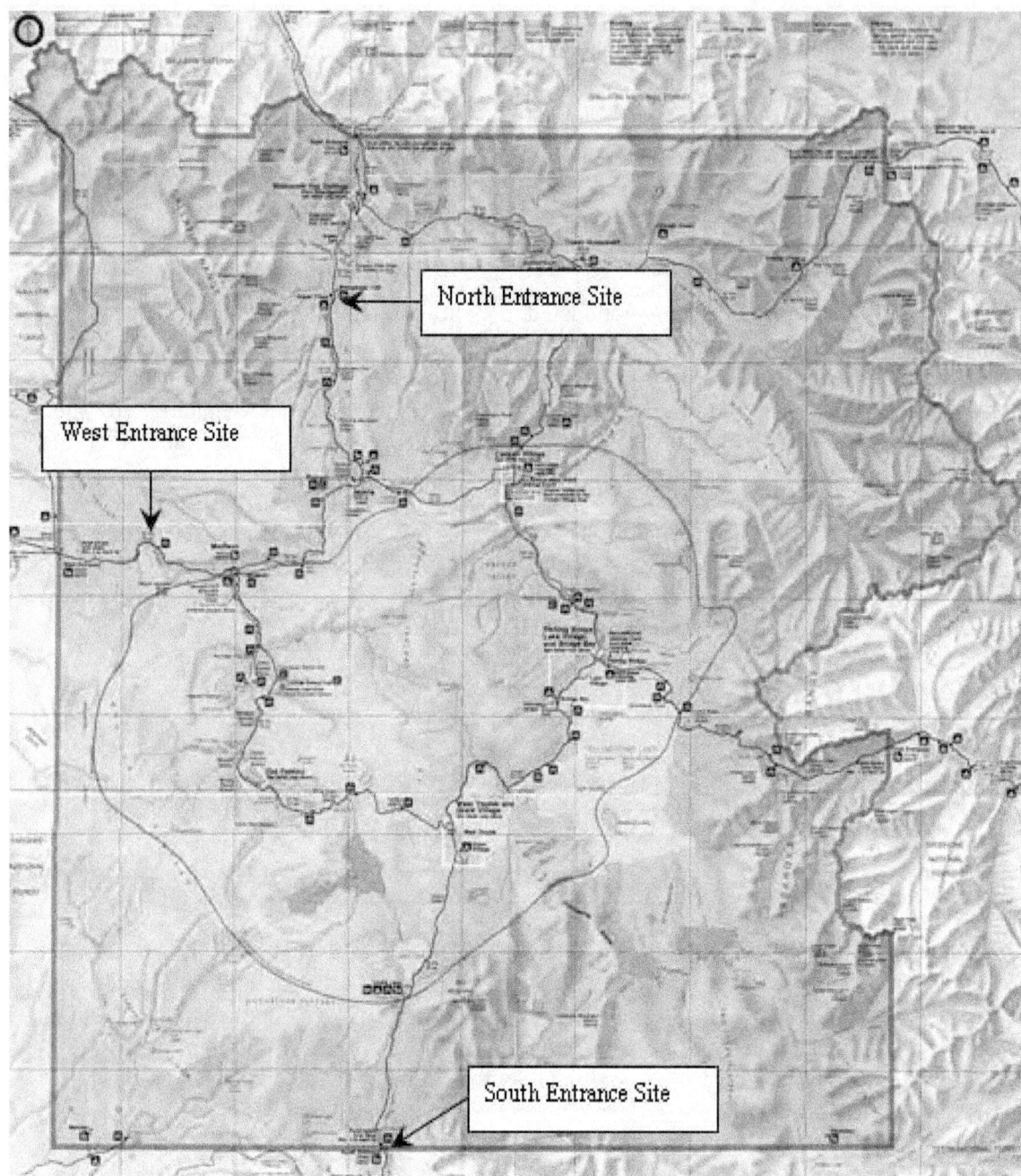

Figure 4. Measurement Site Locations in Yellowstone National Park.

2.1.1 South Entrance Site Location

The site chosen near the south entrance of Yellowstone National Park is shown in the aerial photo in Figure 5 was the same site used during the 2008 Volpe Study and in previous work in 2002[15]. The river shown in Figure 5 was in a ravine approximately 25 feet below the surrounding ground surfaces. Flowing water could be faintly heard during the quietest times, however, it could not be heard during measurements. To the east of the track there was a pull-off where tourists would stop to take photographs. The western pull-off was not present during the measurements due to snow cover. To the north there were several park buildings which did not interfere with the measurements.

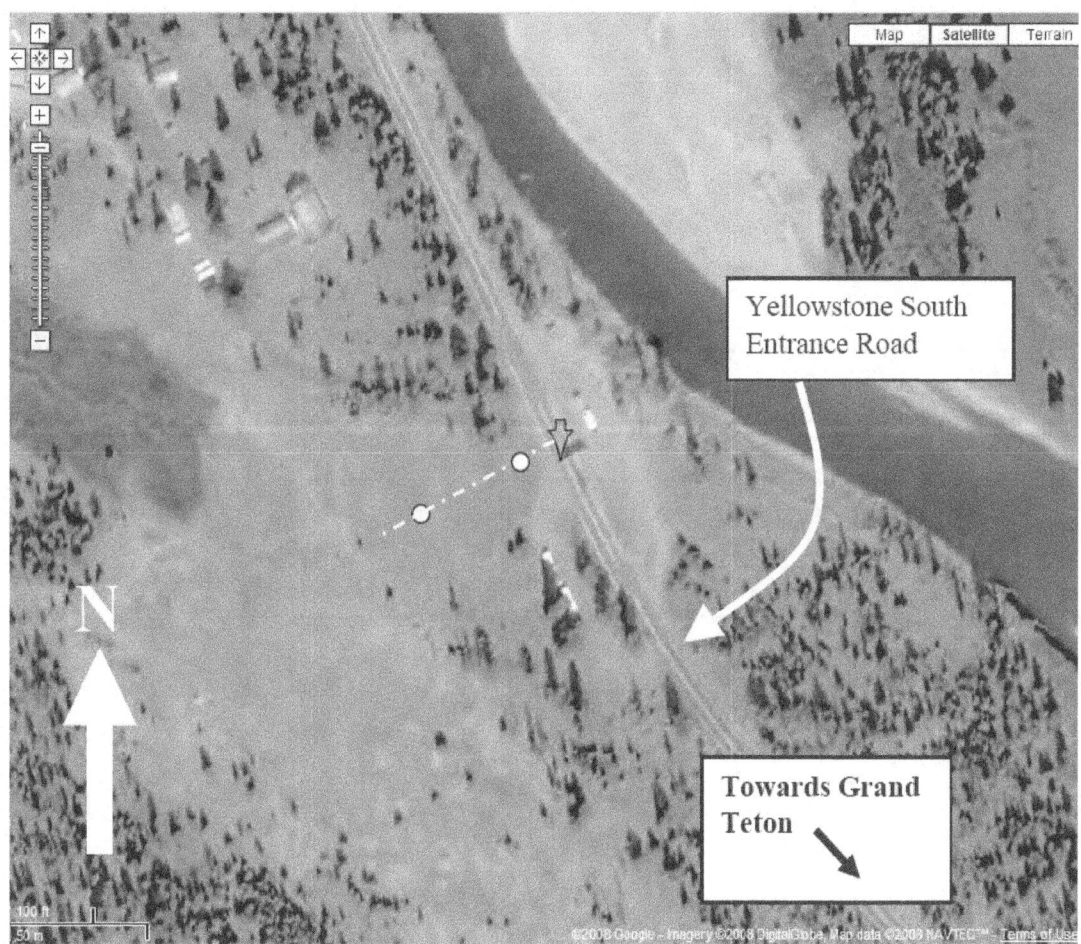

Figure 5. Aerial Photo of the South Measurement Site (44.133, -110.66503)[*]. White dots indicate approximate location of microphones.

Microphones were set up to the west of the track in the region free of trees. A sketch of the site showing the aerial view and profile is provided in Figure 6. The location of equipment is indicated in Figure 6A. Over the course of the winter, snow on the track was groomed and packed while snow off the track was not managed; thus, the snow at the microphone locations was approximately 1 to 2 feet higher than the snow along the travel lane at the measurement site. This resulted in portions of the snowcoach noise sources being at least partially blocked along a line of sight (LOS) to the microphone. An example of this LOS blockage can be seen in Figure 7 where the snowcoach tracks are occluded by the snow cover[†]. This occlusion deviates from the

---

[*] Google Maps, 25 March 08. (http://maps.google.com/maps?f=q&hl=en&geocode=&q=44.133,+-110.66503&ie=UTF8&ll=44.132942,-110.665069&spn=0.002861,0.004989&t=h&z=18)

[†] The LOS is slightly different at the microphone because it is closer but also lower than the camera location, however, this photograph does illustrate the issue.

guidance in the SAE standards. Unfortunately, this could not be remedied. Section 4 discusses the implications of the occlusion in more detail.

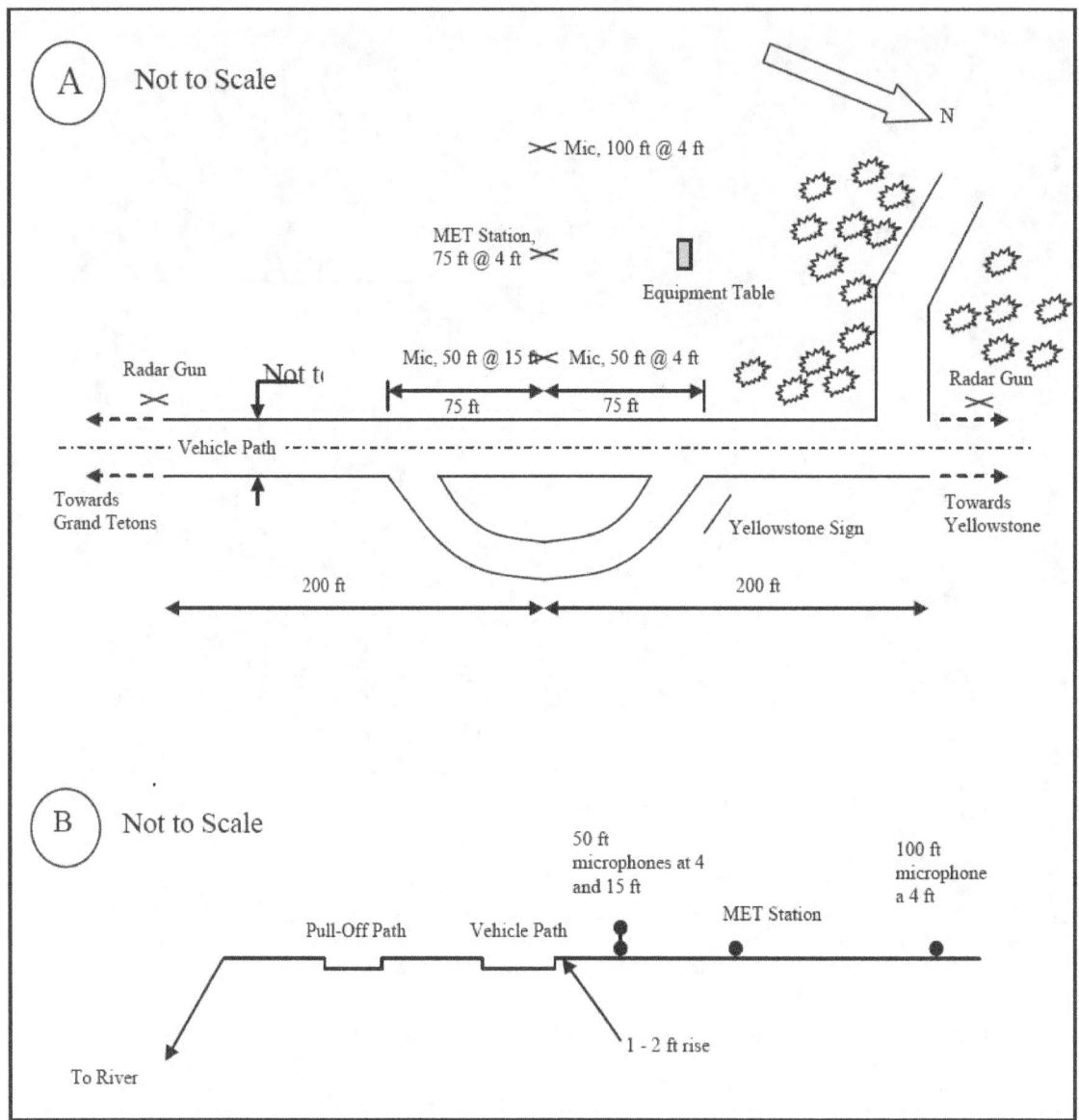

Figure 6. South Measurement Site Sketch – (A) Aerial View, (B) Profile (Not to Scale)

Aside from the difference in snow height between the track and the microphone locations, the only other obstruction was the presence of evergreen trees on the north end of the measurement area. These trees were dispersed and, when listening to pass-by events, they were not observed to affect the sound levels at the 50 foot microphone.

Figure 7. South Entrance Line-of-Sight from Microphone to Tracks

2.1.2 North Entrance Site Location

The site chosen near the north entrance of Yellowstone National Park can be seen in the aerial photo in Figure 8. There were some trees on the east side of the test area as well as to the south west which did not interfere with the measurements. The river, shown in Figure 8, was approximately 400 feet away and was not audible from the test site. To the west of the site there was a warming hut for skiers. Measurements were interrupted when the skier's snowcoaches drove through the test site.

Microphones were set up to the west of the track in the region free of trees. A sketch of the site showing the aerial view and profile is provided in Figure 9. The location of the equipment is indicated in Figure 9A.

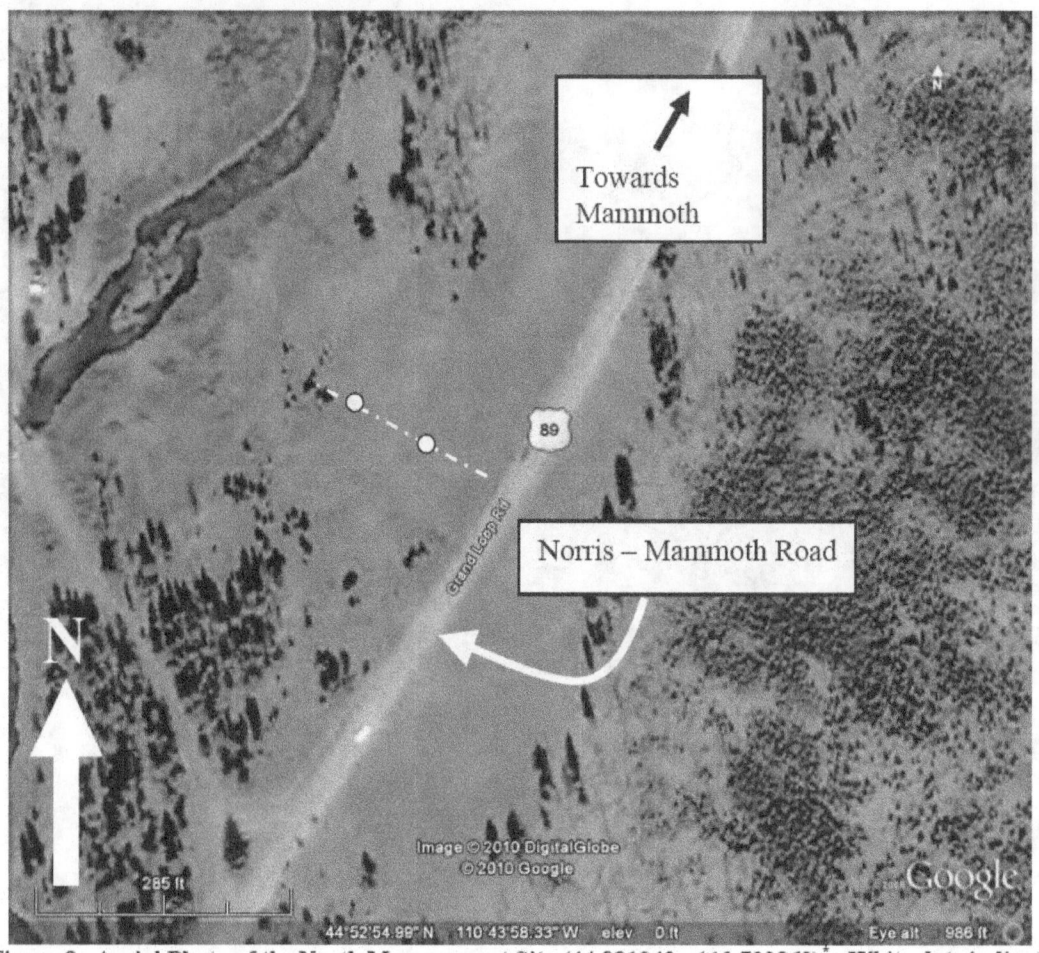

Figure 8. Aerial Photo of the North Measurement Site (44.881942, -110.732869)[*]. White dots indicate approximate location of microphones.

---

[*] Google Maps, 25 March 08. (http://maps.google.com/maps?f=q&source=s_q&hl=en&geocode=&q=44.881942,+-110.732869&sll=44.666951,-110.969993&sspn=0.003502,0.006491&ie=UTF8&ll=44.883151,-110.732875&spn=0.006979,0.012982&t=h&z=16)

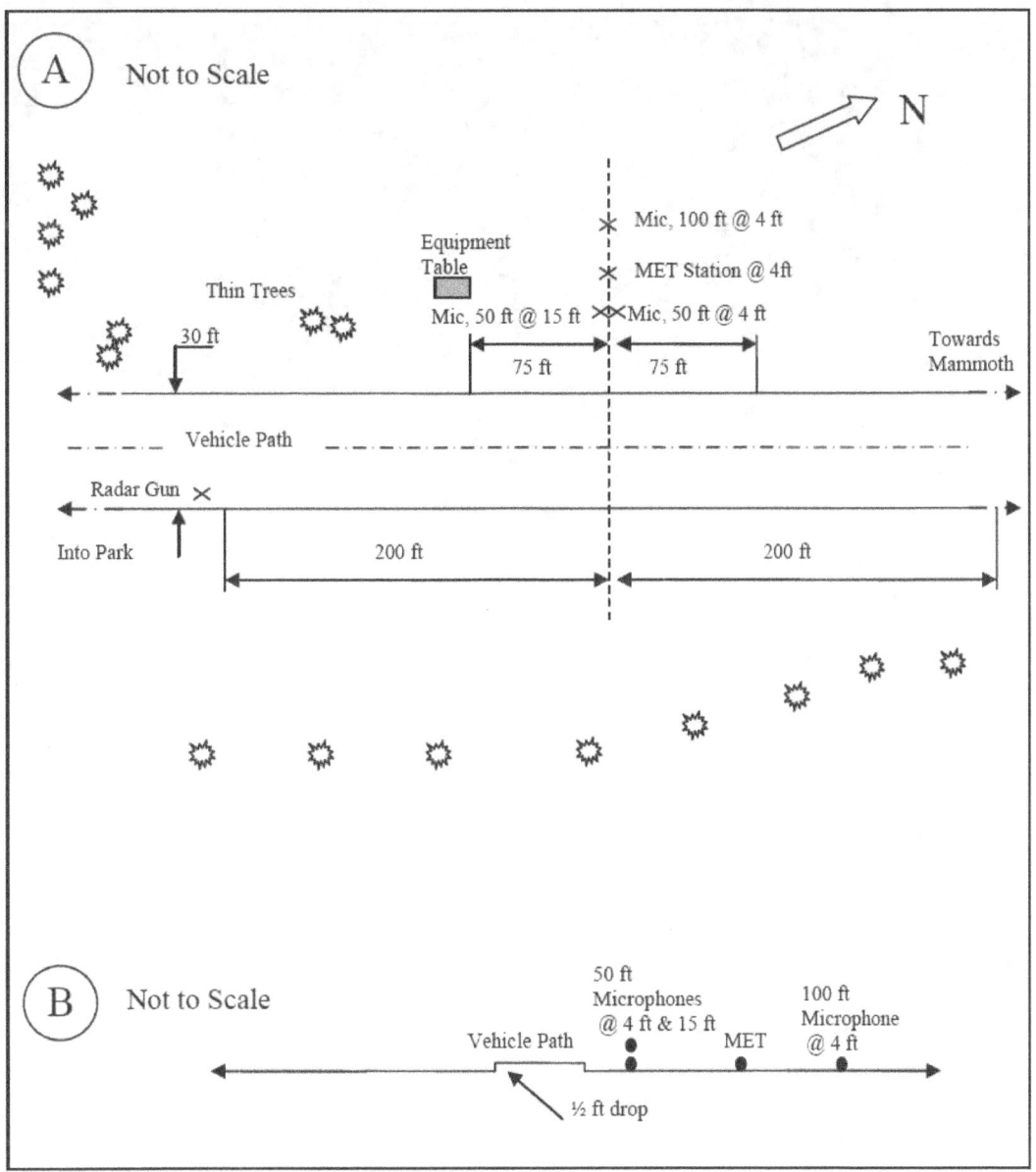

Figure 9. North Measurement Site Sketch – (A) Aerial View, (B) Profile (Not to Scale)

The track surface at the north entrance is elevated about one half foot relative to the adjacent snow cover. This allowed for a clean LOS between the microphone and snowcoach. A view of the LOS from microphone to snowcoach is presented in Figure 10.

Figure 10. North Entrance Line-of-Sight from Microphone to Tracks

2.1.3 West Entrance Site Location

The site chosen near the west entrance to Yellowstone National Park can be seen in the aerial photo in Figure 11. Trees to the west of the track are approximately 230 feet from the center of the vehicle path. To the east there is a river that was approximately 100 feet from the center of the vehicle path. The river sits approximately 5 feet below the vehicle path with a few scattered evergreen trees between the vehicle path and river. Beyond the river, on the opposite side of the vehicle path as the microphone array, there was a steep hill that was covered with scattered evergreen trees. Flowing water could be heard during the quietest periods but not during measurements.

Microphones were set up to the south-west of the track in the region free of trees. A sketch of the site showing the aerial view and profile is provided in Figure 12. The location of the equipment is indicated in Figure 12A.

Figure 11. Aerial Photo of the West Measurement Site (44.6667, -110.9702)[*]. White dots indicate approximate location of microphones.

---

[*] Google Maps, 25 March 08. (http://maps.google.com/maps?f=q&source=s_q&hl=en&geocode=&q=44.6667,+-110.9702&sll=44.66612,-110.969998&sspn=0.001751,0.003245&ie=UTF8&ll=44.666951,-110.969993&spn=0.003502,0.006491&t=h&z=17)

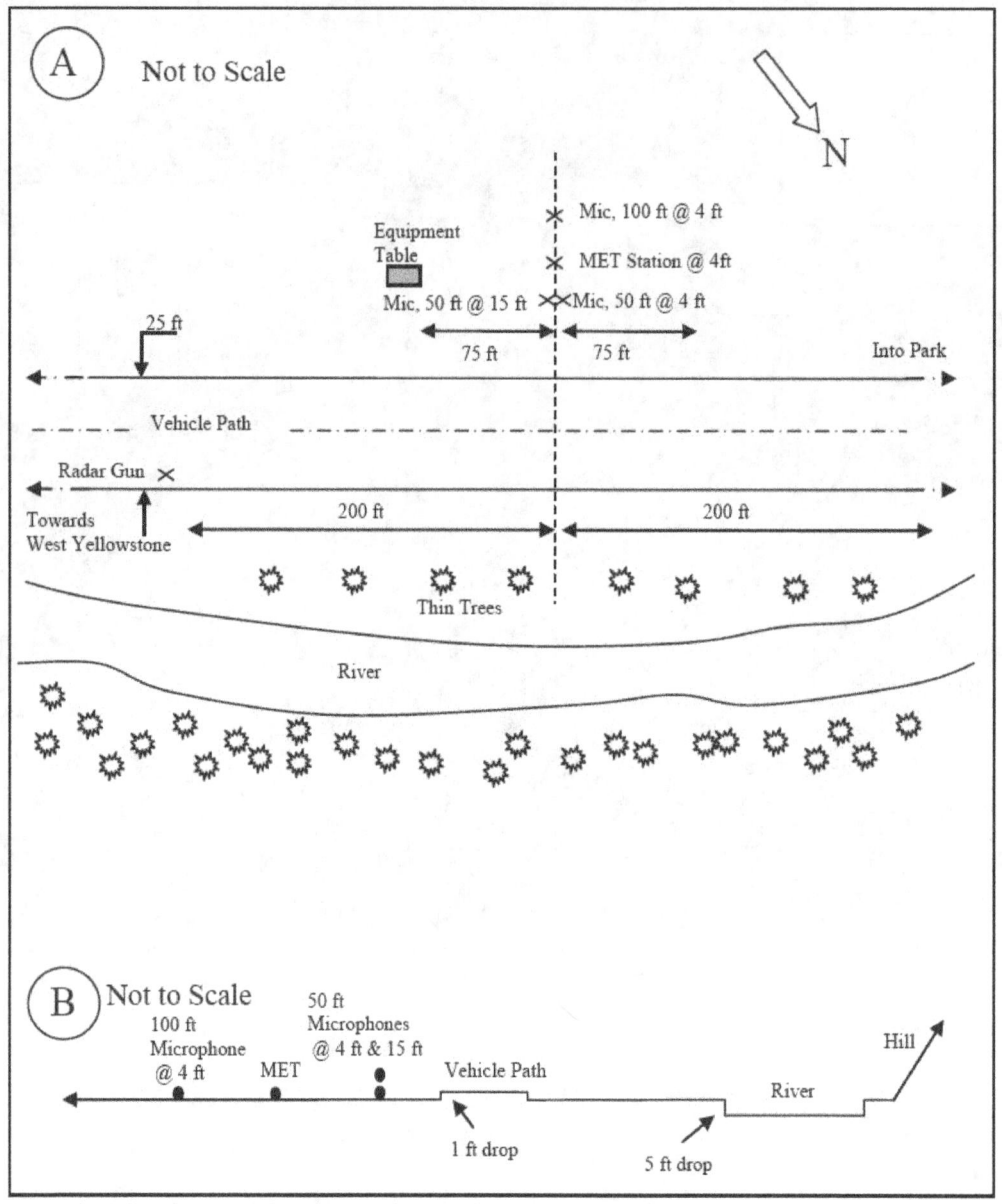

Figure 12. West Measurement Site Sketch – (A) Aerial View, (B) Profile (Not to Scale)

The track surface at the west entrance is elevated about 1 foot relative to the adjacent snow cover. This allowed for an unobstructed LOS between the microphone and snowcoach. A view of the LOS from microphone to snowcoach can be seen in Figure 13.

There were 2 obstructions to note at this site. The first obstruction is the scattered evergreen trees on the northeast side of the vehicle path. These did not affect the measured data. The second obstruction to note is the hill on the northeast side approximately 320 feet from the vehicle path. Due to the hill being covered with acoustically absorptive snow and scattered evergreen trees, reflection of sound can be neglected.

Figure 13. West Entrance Line-of-Sight from Microphone to Tracks

2.2 Measurement Conditions

None of the roads in the vicinity of the three measurement sites were closed during the measurements so extraneous road traffic did occur, primarily during the early morning and late afternoon. Extraneous traffic predominantly consisted of guided tour groups, either on snowmobile or in snowcoaches that entered Yellowstone in the morning and departed in the afternoon. This necessitated a pause during the measurements whenever a group came through. Temperatures during the measurements ranged from 12 to 37 degrees Fahrenheit. Relative humidity ranged from 26 to 83 percent. Wind speeds ranged from 0 to 17 mph, although acoustic measurements made during wind speeds greater than 12 mph were discarded, in accordance with SAE J1161.

2.3 Vehicle Description

Over the course of the six-day study, the team measured each side of 25 snowcoaches. The vehicles included 17 unique models, both at two constant pass-by speeds and idle. A summary of the snowcoach manufacturer and model type, track type, and engine characteristics is provided in Table 8. Photos of the vehicles are shown in Figure 14 through Figure 30. The Alpen Guide (Kitty), Xanterra 713, Yellowstone Snowcoach Tours (SNOVAN5), and Yellowstone Expedition (Hayden) shown in Figure 14 though Figure 17 were tested at all three sites.

Table 8. Snowcoach Vehicle Description

Snowcoach	Designation	Manufacturer	Model	Track Type*	Engine Year	Engine Size	Fuel Type	Passenger Capacity
Alpen Guide**	Kitty	Bombardier	B-12	Tracks & Skis	2002	5.3 L V-8 Chevy Vortek fuel-injected	Gasoline	10
Buffalo Bus Touring	#3	Ford	E-350 Comm-Trans	Mattracks	2006	6..9 L V-10	Gasoline	14
Buffalo Bus Touring***	#4	Ford	F550 "Krystal"	Grip Tracks	2009	6.4 L 8 Cylinder	Diesel	32
Buffalo Bus Touring	T2	Ford	E-350 "Vanterra"	Mattracks	2005	6..9 L V-10	Gasoline	14
See Yellowstone Tours	#4	Ford	Econoline	Mattracks 150 with skis	2000	6.8 L V-10	Gasoline	9
See Yellowstone Tours	#6	Ford	Vanterra	Mattracks 150 with skis	2004	6.8L V-10	Gasoline	13
See Yellowstone Tours	#9	Ford	Odyssey	Tank Tracks with skis	2007	6.0 L	Diesel	20
Xanterra	430	Chevrolet	Express Van	Mattracks	2008	6.0 L V-8 GMC	Gasoline	12
Xanterra**	707	Bombardier	B-12	Tracks & Skis	2006	5.7 L V-8 GMC	Gasoline	10
Xanterra	709	Bombardier	B-12	Tracks & Skis	2006	5.7 L V-8 GMC	Gasoline	10
Xanterra**	710	Bombardier	B-12	Tracks & Skis	2006	5.7 L V-8 GMC	Gasoline	10
Xanterra**	713	Bombardier	B-12	Tracks & Skis	2006	5.7 L V-8 GMC	Gasoline	10
Xanterra	537	Pirnoth	Powder Tour Cat TR	Rubber/Pirnoth	2008	5.2 L V-8 Dodge	Gasoline	15
Yellowstone Expedition	Hayden	Dodge	B350 Van	Snowbuster wide track	1994	5.2 L V-8	Gasoline	10
Yellowstone Expedition	Eleanor	Ford	E150 Van	Snowbuster – Bombardier tracks	1997	4.6 L V-8	Gasoline	8
Yellowstone Snowcoach	SNOVAN4	Ford	Econoline	Mattracks	2000	6.8 L V-10	Gasoline	10
Yellowstone Snowcoach	SNOVAN5	Ford	Econoline	Mattracks	2001	6.8 L V-10	Gasoline	10

* Track types for the snowcoaches are as follows: Skis / Tracks – the rear wheels have been replaced with tracks while the front wheels have been replaced with skis; Mattracks – both the front and rear wheels have been replaced with triangular tracks manufactured by Mattracks; Snowbuster – are a specific manufacturer of Skis / Tracks, and Grip Trax – another specific type of triangular tracks to replace the front and rear wheels.

** These vehicles have been retrofitted for a reduced sound level.

*** Vehicle includes an air emission reduction device which operates periodically and creates additional noise. The device did not operate during testing, however at other times the pollution-control device was clearly audible. It is recommended that future testing endeavor to include these types of sound sources.

Figure 14. Purpose Built Bombardier B-12 with Tracks and Skis (Alpen Guide – Kitty)

Figure 15. Converted Ford E-350 Comm-Trans van with Mattracks (Buffalo Bus Touring #3)

Figure 16. Converted Ford F550 "Krystal" van with Grip Tracks (Buffalo Bus Touring – #4)

Figure 17. Converted Ford E-350 Vanterra van with Mattracks (Buffalo Bus Touring – T2)

Figure 18. Converted Ford Econoline van with Mattracks 150 (See Yellowstone Tours – #4)

Figure 19. Converted Ford Vanterra with Mattracks 150 (See Yellowstone Tours – #6)

Figure 20. Converted Ford Odyssey van with Mattracks (See Yellowstone Tours – #9)

Figure 21: Converted Chevrolet Express Van with Mattracks (Xanterra – 430)

Figure 22. Purpose Built Bombardier B-12 with Tracks and Skis (Xanterra – 707[*])

Figure 23. Purpose Built Bombardier B-12 with Tracks and Skis (Xanterra 709*)

---

[*] It should be noted that the Xanterra models 707, 709, 710, and 713 were equipped with a tarp on the top to cover the storage rack. This tarp did not produce any noticeable noise, i.e. including flapping, during the measurements.

USDOT Research & Innovative Technology Administration
Environmental Measurement and Modeling Division
Volpe Center Acoustics Facility

April 2010

Figure 24. Purpose Built Bombardier B-12 with Tracks and Skis (Xanterra – 710[*])

Figure 25. Purpose Built Bombardier B-12 with Tracks and Skis (Xanterra – 713[*])

---

[*] It should be noted that the Xanterra models 707, 709, 710, and 713 were equipped with a tarp on the top to cover the storage rack. This tarp did not produce any noticeable noise, i.e. including flapping, during the measurements.

Figure 26. Purpose Built Pirnoth Powder Tour Cat TR with Rubber/Tracks (Xanterra Pirnoth – 537)

Figure 27. Converted Dodge B350 van with Snowbusters (Yellowstone Expeditions – Hayden)

---

* It should be noted that the Xanterra models 707, 709, 710, and 713 were equipped with a tarp on the top to cover the storage rack. This tarp did not produce any noticeable noise, i.e. including flapping, during the measurements.

Figure 28. Converted Ford E150 van with Snowbuster (Yellowstone Expeditions – Eleanor)

Figure 29. Converted Ford Econoline van with Mattracks (Yellowstone Snowcoach Tours – SNOWVAN4)

Figure 30. Converted Ford Econoline van with Mattracks (Yellowstone Snowcoach Tours – SNOWVAN5)

2.4 Equipment Description

Three acoustic systems, one meteorological system, and a radar-based speed detector were used to collect the requisite data at each site.

2.4.1 Acoustic System

Each acoustic system consisted of a Larson Davis Model 824 Real Time Analyzer (LD824), a Bruel & Kjaer (B&K) Model 4189 1/2-inch pressure-response electret microphone and a Sony D100 Digital Audio Tape (DAT) recorder as backup to the LD824. Figure 31 shows a block diagram of the acoustic system. The three microphone/preamp systems were mounted on tripods at a height of 4 or 15 feet above the snow cover and connected using extension cables to the LD824 positioned at the measurement table, so that all acoustic systems could be monitored at a single location. Care was taken to ensure that no connections were exposed to moisture. This involved using plastic bags to cover connections and, wherever possible, keeping cables off the snow.

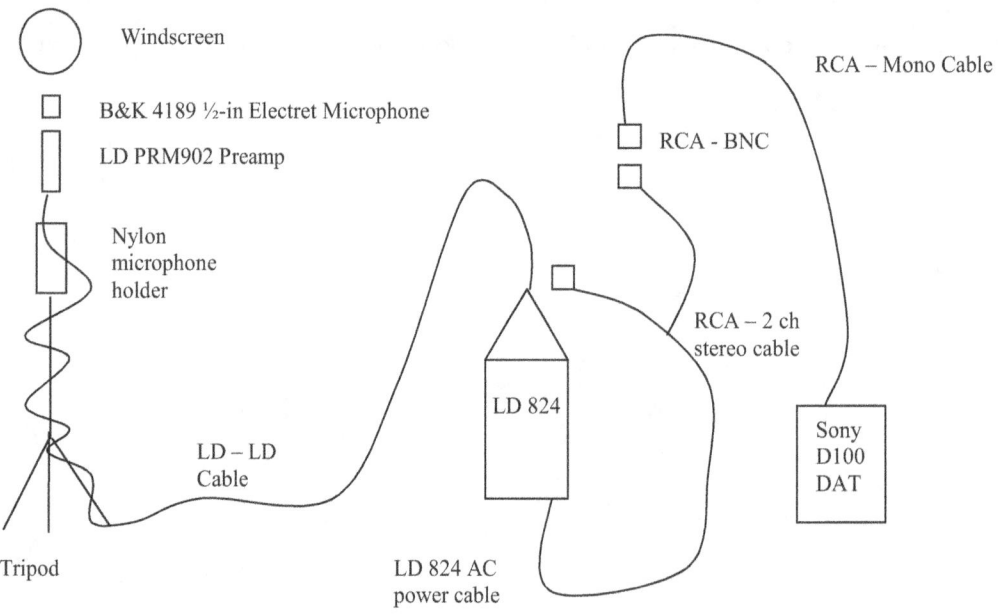

Figure 31. Acoustic System Setup

One-second A-weighted equivalent sound level (L_{Aeq}) and the maximum A-weighted sound level (L_{ASmx}) data were stored in the LD824s, along with one-third octave band sound levels ranging from 12.5 Hz to 20 kHz for each record. Information on the settings used for the LD824 and the Sony D100 DAT is provided in Appendices A and B.

Because the temperatures during the measurements were expected to be below some of the measurement instrumentation manufacturer's recommendation for operating temperature, special considerations were taken, see Table 9 for operating temperatures. The sound level meters and recorders were placed atop hand warmers in a case fitted with insulating foam (see Figure 32).

Table 9. Instrumentation Operating Temperatures.

Instrument	Temperature °F	
	Minimum	Maximum
LD PRM902 Preamp	-40	149
B&K 4189 Microphone	-22	257
LD 824	14	122

Figure 32. Instrumentation Case

2.4.2 Meteorological Equipment

A Qualimetrics Transportable Automated Meteorological Station (TAMS) was used to measure wind speed, wind direction, relative humidity, barometric pressure, and air temperature at one-second intervals. A complete TAMS system consists of a sensor unit and a control / display unit that displays real time meteorological data. These battery-powered stations are portable and well suited for remote sampling. Table 10 shows the metrics collected and the operating accuracy, Figure 33 presents a block diagram of the system. The unit's sensors were placed on a tripod at a height of 4 feet above the snow cover. Data from the control unit is automatically displayed and saved onto a Hewlett Packard (HP) 200LX Palmtop hand held computer.

Table 10. TAMS Measurement Capabilities

Metric	Units	Range	Resolution	Accuracy
Wind Speed	Miles Per Hour (mph)	1 to 55 mph	1 mph	1 mph or 5% of range
Wind direction	Degrees (°)	360°	10°	root mean standard error of 18°
Temperature	Degrees Fahrenheit (°F)	-9 to 110 °F	1 °F	1 °F
Relative Humidity	Percent (%)	0 to 100 %	1%	3%
Barometric Pressure	Millibars (mb)	610 to 1084 mb	.1 mb	1 mb

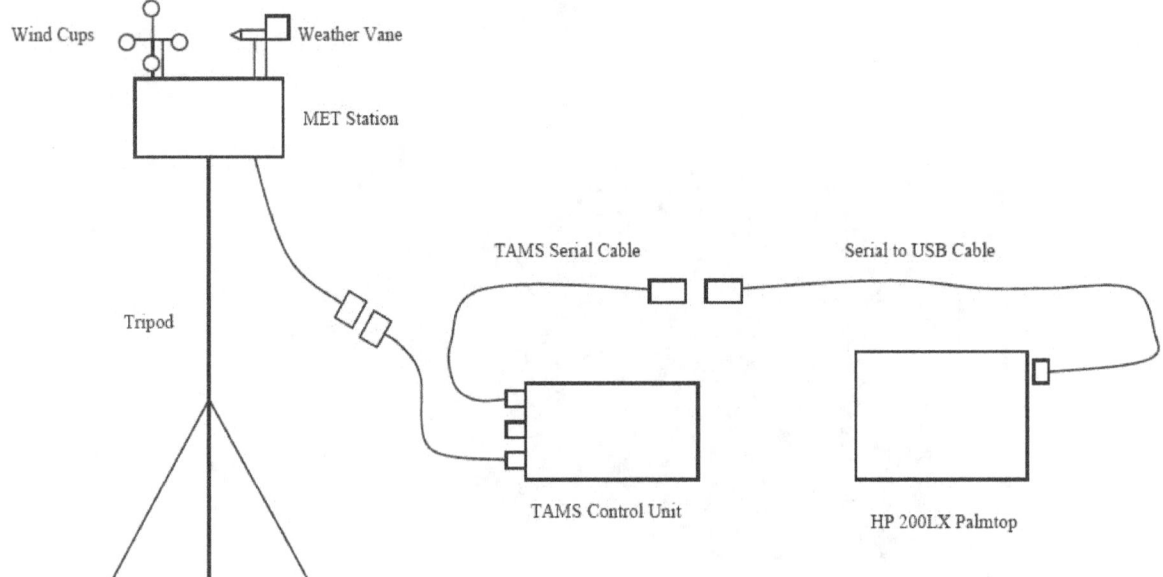

Figure 33. TAMS Setup

2.4.3 Vehicle Speed Collection

A Doppler-radar based speed detector was used to measure vehicle speed for each event. It was operated at one end of the vehicle travel path. The location for each site is indicated in Figure 6, Figure 9, and Figure 12 for the south, north, and west entrance sites, respectively. In addition, a handheld GPS unit was used in the vehicle to assist the driver in maintaining a constant speed.

3 Measurement Protocol

This section details the measurement protocol used with specific references to SAE J1161. The measurements were conducted jointly by the NPS and Volpe. The NPS contracted the use of the vehicles, determined the measurement sites, and provided tools to help snowcoach drivers control their speed during the measurements. The NPS also provided spare acoustic instrumentation, as well as staff to assist as needed. Volpe provided acoustic, meteorological, and speed measurement equipment (see Section 2.4) and staff to conduct the measurements. Details of the measurement protocol and a sample log sheet are provided in Appendix C.

3.1 Methodology in Accordance with SAE International Standard J1161

SAE J1161 specifies a methodology for the measurement of exterior operational sound levels for snowmobiles. This methodology has been used in this study as snowmobiles and OSV's have similar propulsion systems, namely tracks and skis and are used at similar speeds in the parks. The measurement area is required to be an open region of packed snow at least 2 inches deep which is free of reflecting surfaces. The snowpack is required to be sufficient to support the OSV without penetration and with no more than 3 inches of loose snow on top of the packed snow. An illustration of measurement area requirements is provided in Figure 34.

Per SAE J1161, the OSV approaches the measurement area at a speed of 15 mph and maintains this speed throughout the area. The microphone / sound level meter is required to be positioned 50 feet from the center of the travel lane and 4 feet above the snow cover (see Figure 34). The sound level meter should be set to measure the maximum A-weighted sound pressure level with slow time weighting (L_{ASmx}) during vehicle pass-bys at a constant speed of 15 mph. Measurements are repeated until three L_{ASmx} values are measured within 2 dB. These L_{ASmx} values are then arithmetically averaged and rounded to the nearest integer. The measurements are conducted for the vehicle traveling in both directions and the side of the vehicle with the highest average is reported as the sound level for the OSV. The sides of the vehicle are designated as the "right side" and "left side" of the vehicle, from the perspective of the vehicles driver. An example of the sound level calculation is provided in Table 11.

The standard specifies that, during the measurements, the atmospheric temperature, pressure, relative humidity, wind speed and direction be measured at recorded intervals of not less than 1 hour. Further details are provided in Reference 14.

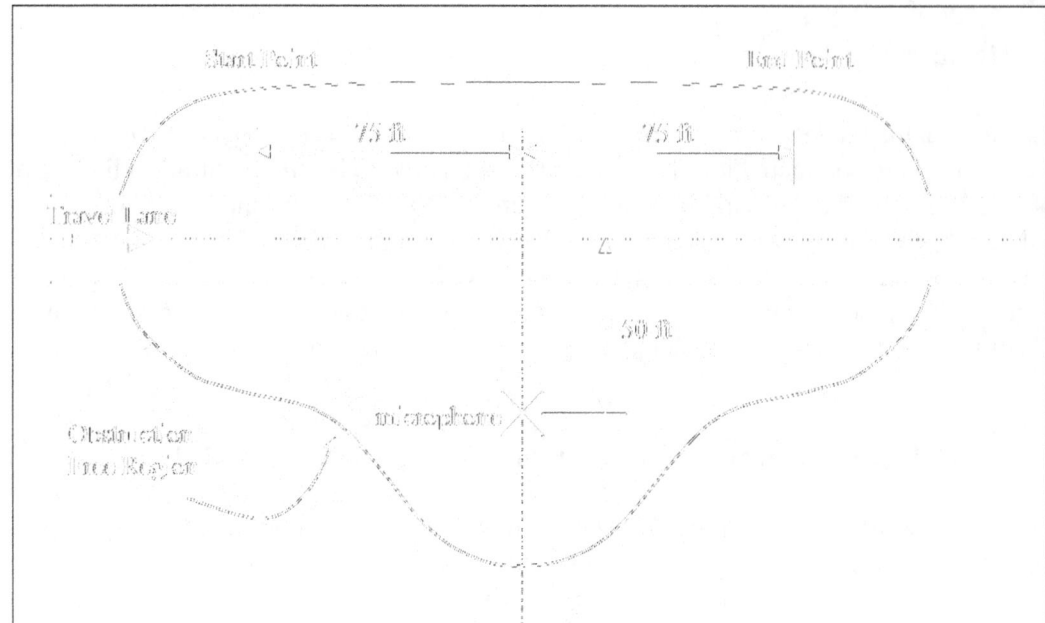

Figure 34. SAE J1161 Measurement Area Layout

Table 11. Example Sound Level Calculation for SAE J1161

Pass-by	L_{ASmx} Sound Level (dBA)			Within 2 dB?	Average	Final Sound Level
Left Side	64	65	65	yes	65	65
Right Side	62	61	62	yes	62	

3.2 Measurements Supplemental to SAE J1161 Methodology

In order to maximize the data obtained during the study, vehicles were operated under additional operating conditions, additional microphones were utilized, and the measurements were made over an extended measurement area. In general, increasing the load of a snowcoach will increase its sound level. Two loading factors could affect the test conditions:

1. The number of passengers;
2. The snow conditions on the track.

The snowcoaches tested had carrying capacities which ranged from eight to thirty-two passengers. It was not practical to test all snowcoaches at full capacity. For consistency, generally only the driver and one passenger were used for the measurement runs. Thus, the sound levels measured represent the lowest sound levels that these vehicles would produce with respect to passenger load. Although the absolute sound levels during loaded operations will likely be higher, unloaded comparisons can be used to evaluate relative comparisons between vehicles.

During the testing of multiple snowcoaches in a day, the snow on the vehicle path became softer and provided less traction and more drag, requiring snowcoaches to operate at higher engine speeds to obtain the same vehicle speed. In order to reduce this effect, a groomer was used to groom the track periodically and at night. Grooming during the day was at the discretion of the snowcoach drivers and NPS staff.

The measurement area was extended to cover a region from 200 feet on either side of the microphone line instead of the 75 feet on either side specified by SAE S1161 (see Figure 35). The extended area allowed for an additional microphone to be used in the study. This microphone was located 100 feet from the travel lane at a height of 4 feet above the snow cover. This microphone provided a second measurement near the ground so that comparisons could be made with the 50 foot microphone in order to evaluate the over-ground attenuation at the site. The extension to 200 feet on either side provided sufficient distance to assure a 10 dB drop-off from the maximum sound level, so that Sound Exposure Levels (SEL's) for each snowcoach could also be readily computed.

A third microphone was positioned 50 feet from the center line of travel lane at a height of 15 feet above the snow cover. This microphone provided a measurement that was less influenced by the ground cover and any LOS occlusions between the microphone and the vehicle's track/snow interface.

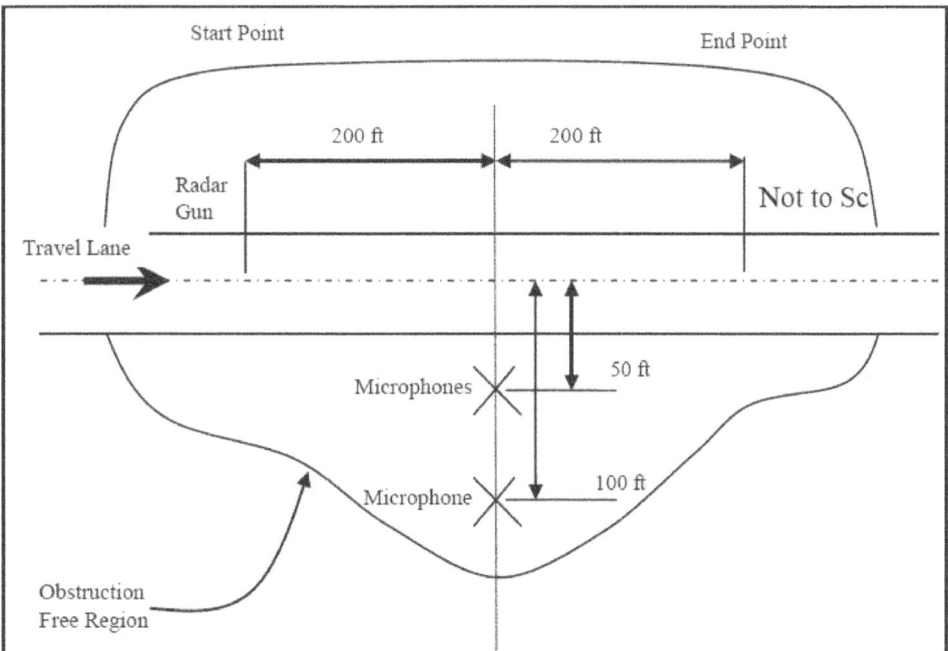

Figure 35. Extended Measurement Area Layout

High constant speed and idle operating conditions were also measured. One minute L_{Aeq} idle measurements were made for each side of the vehicle. High constant-speed, vehicle-dependent, tests were undertaken at the highest constant speed that could be maintained safely through the measurement site, typically about 30 mph.

3.3 Meteorological Measurements

Meteorological measurements were made throughout the measurement period and recorded at 1 second intervals. This is a higher temporal resolution than required by the SAE J1161, but allowed for increased quality control of the measurements, especially with respect to wind speed. Temperature, relative humidity, wind speed, wind direction, and barometric pressure were measured by using a TAMS system located 75 feet from the center of the travel lane, in line with the microphones. These data were displayed and recorded on an HP 200LX palmtop computer, as discussed in Section 2.4.2. The wind speed data was checked to ensure results were not generated from measurements where the wind speed was greater than 12 mph, in accordance with SAE J1161.

4 Results and Analysis

Processed results and analysis are presented in this section. Analysis focuses on the five stated objectives from Section 1:

1. Refine and finalize snowcoach sound level testing procedures recommended from the 2008 Volpe Study[5];
2. Determine which snowcoaches meet the BAT with respect to sound level;
3. Determine any site-specific measurement bias due to variables such as line of sight blockage, snow depth, barometric pressure associated with differing altitudes, etc.;
4. Develop a sound level versus speed relationship to determine if any speed restriction on snowcoaches is needed to meet BAT requirements; and
5. Develop OSV noise-distance relationships for use in the modified version of FAA's Integrated Noise Model (INM) developed for NPS in 2006 Volpe Study[5]. In the 2006 study, the ground-to-ground propagation algorithms in INM Version 6.2[6,7] were modified to better account for sound propagation over snow-covered terrain. The modifications were based on the physical acoustics algorithms in the Federal Highway Administration's (FHWA) Traffic Noise Model (TNM)[8].

4.1 Meteorological Conditions

A summary of the meteorological data is provided in Table 12. The primary concern during the measurements was wind speed. Although recorded max wind speeds exceeded 12 mph on both January 15th and the 16th, care was taken to avoid making measurements during high wind periods.

Table 12. Summary of Meteorological Conditions During the Measurements

Entrance	Date	Temperature (°F)			Relative Humidity (%)			Wind Speed (mph)			Barometric Pressure (millibars)		
		Min.	Max.	Avg.	Min.	Max.	Avg.	Min.	Max.	Avg.	Min.	Max.	Avg.
South	1/14/2009	26.8	32.0	28.7	67	79	73	0.0	8.5	2.4	796	797	797
North	1/15/2009	25.2	33.3	30.6	60	82	66	2.0	15.0	6.9	786	787	787
	1/16/2009	20.5	35.4	30.0	42	71	61	0.0	16.8	6.8	787	790	788
West	1/20/2009	21.9	36.5	30.9	34	61	42	0.0	6.0	3.1	803	806	804
	1/21/2009	12.6	36.5	27.5	26	79	47	0.0	8.5	2.8	796	800	798
	1/22/2009	16.0	31.6	24.8	60	83	71	0.0	9.8	3.7	793	794	794

4.2 Sound Level Results

This section presents the sound level results for each snowcoach, including pass-by sound level time histories (section 4.2.2), overall Maximum Sound Levels (L_{ASmx})(section 4.2.3), and Sound Exposure Levels (SEL) (section 4.2.5), for low speed, high speed, and idle events. It also presents the relationship between sound level and speed (section 4.2.4). Maximum spectra for low speed and high speed events are discussed in section 4.2.6.

4.2.1 Ambient Sound Levels

During measurements, the primary sound sources which contributed to the background ambient sound level included OSV's operated at other locations in and outside the parks, birds, and wind / vegetation noise. The range of ambient sound levels for each day is presented in Table 13. The settings used on the LD824 allow levels to be measured down to approximately 23 dBA with a $^+/-$ 1 dBA accuracy[*].

The higher ambient sound levels directly correlated with more human generated sound and higher wind speeds. For the purposes of this study, these estimates of the ambient sound level provide a conservative limit for comparison with vehicle noise measurements with respect to event quality, i.e., source level above ambient. During the low and high speed measurements, the quietest snowcoaches were at least 16 dBA above the loudest ambient. The idle measurements were often within 10 dBA of the ambient sound level; these cases are noted in Table 16.

Table 13. Summary of Ambient Conditions During Measurements

Entrance	Date	Range of Background Ambient Sound Levels (dBA)	
		Min	Max
South	1/14/2009	22	24
North	1/15/2009	32	42
	1/16/2009	28	34
West	1/20/2009	22	24
	1/21/2009	23	35
	1/22/2009	26	33

4.2.2 Sound Level Time History

An example sound level time history for the Yellowstone Snowcoach (SNOVAN4) Gasoline Econoline snowcoach operating at low speed (15 mph) pass-by is shown in Figure 36. The underlying data are 1-second, equivalent sound levels $L_{Aeq,1sec}$. This figure shows the general pattern observed for pass-by events: values increase as the snowcoach approaches the microphone center line; there is a peak region; values then decrease as the snowcoach travels away from the microphone center line. Another important feature to observe is that the peak sound level is at least 10 dBA greater than the lowest approach and lowest departure sound levels. This minimum 10 dBA criterion was maintained throughout the study and ensures that a clean pass-by was acquired and that analyses using these data will not be unduly contaminated with sound sources other than the vehicle under study. An example time history profile is given in Figure 37 for a high speed (33 mph) pass-by of the Yellowstone Snowcoach – SNOVAN4 snowcoach.

---

[*] During data collection +20 dB of gain was applied to the LD824; per manufacturer specifications, this results in an instrument noise floor of 15 dBA. A limitation of a sound level meter is that as sound levels approach the floor the linear response becomes less accurate. In the case of the LD824 with the settings used for this project, below 28 dBA the instrument linearity does not meet ANSI S1.4, American National Standard – Specification for Sound Level Meters. The manufacturer indicated that the LD824 can measure down to 23 dBA with $^+/-$ 1 dB with the settings used.

An example idle measurement time history is shown in Figure 38 for the Yellowstone Snowcoach (SNOVAN4) Gasoline Econoline snowcoach. The profile is different here because the vehicle is stationary during the measurement. Because the sound level does not change significantly over the measurement period it is not readily apparent that this is a clean measurement. In order to verify that these data are not contaminated by extraneous sounds, they are compared with measured ambient sound levels. Example sound level time histories for all vehicles and operating speeds/conditions are provided in Appendix E.

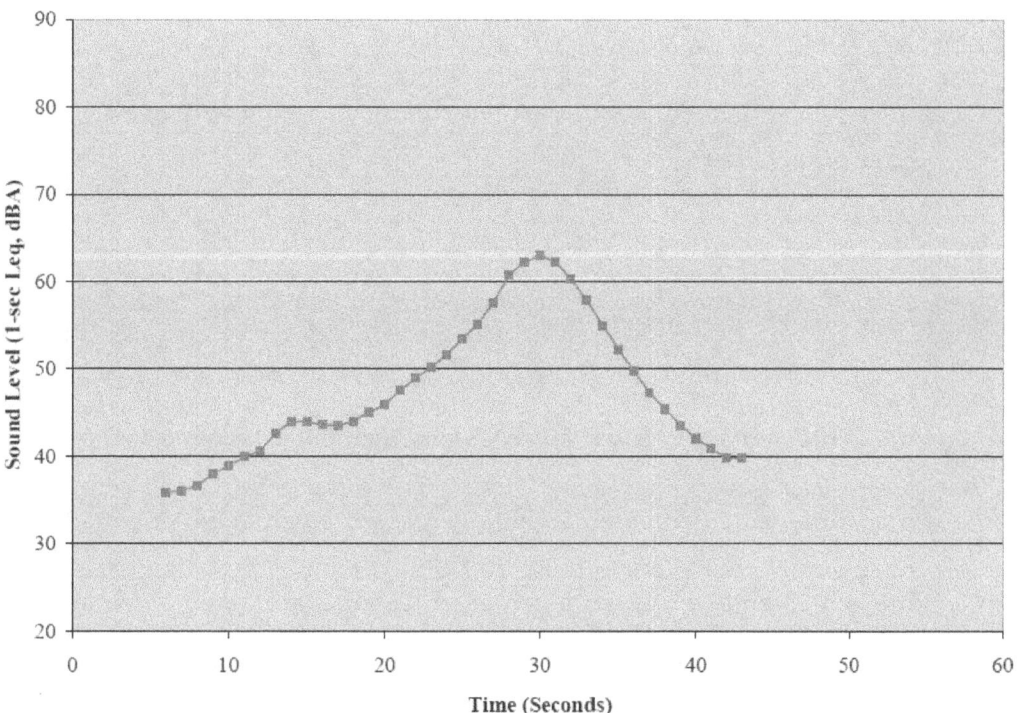

Figure 36. Yellowstone Snowcoach (SNOVAN4) Van, Left Side at Low Speed (15 mph)

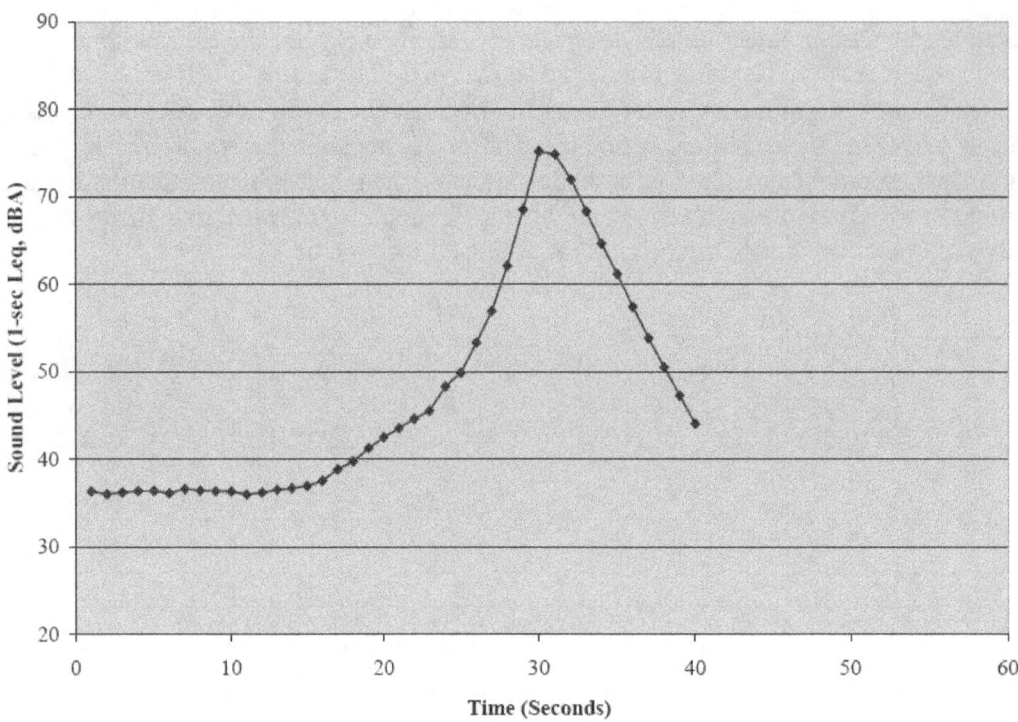

Figure 37. Yellowstone Snowcoach (SNOVAN4) Van, Right Side at High Speed (33 mph)

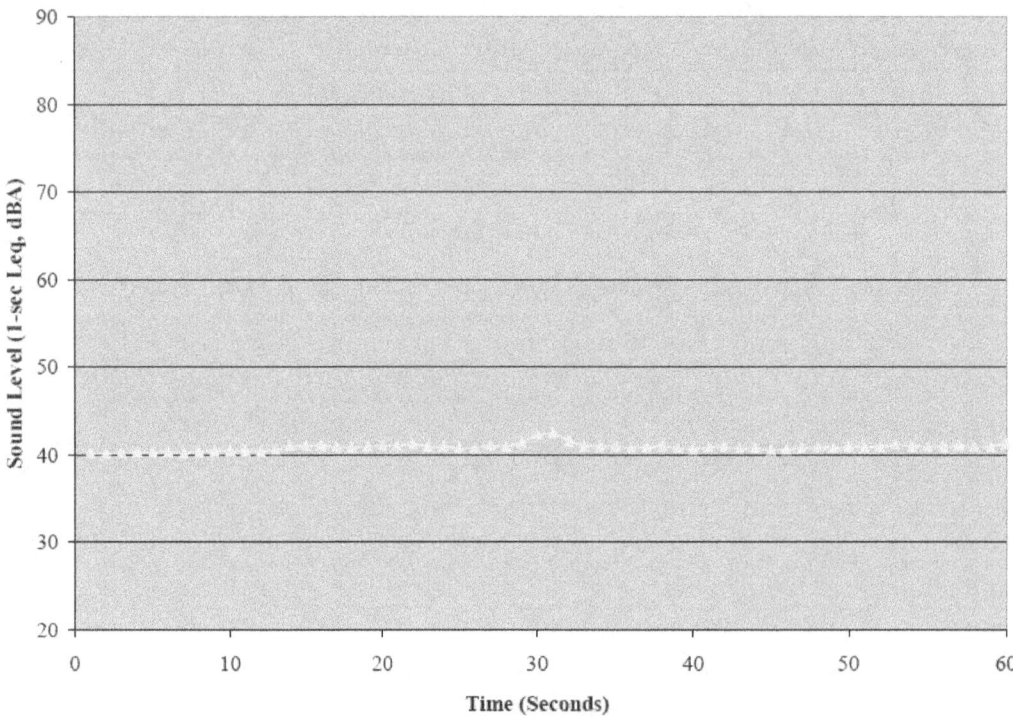

Figure 38. Yellowstone Snowcoach (SNOVAN4) Van, Left Side at Idle

4.2.3 Maximum A-Weighted Sound Levels (L_{ASmx})

In accordance with SAE J1161, pass-by measurements are repeated until three L_{ASmx} values are measured within 2 dB. These L_{ASmx} values are then arithmetically averaged and rounded to the nearest integer. The measurements are conducted for the vehicle traveling in both directions and the side of the vehicle with the highest average is reported as the sound level for the OSV. A summary of all the L_{ASmx} values that were used to produce the reported maximum sound levels is provided in Appendix E.

Table 14 rank orders the twenty five snowcoaches measured from highest to lowest maximum A-weighted sound pressure levels with slow time weighting (L_{ASmx}) for constant low speed pass-bys [15 mph (nominal)] for the 50 foot microphone. The values listed in Table 14 are the arithmetically averaged L_{ASmx} values for the pass-by direction which yielded the highest average value. Averages of the measured speeds are specified in the last column. Table 15 rank orders the twenty five snowcoaches measured from highest to lowest based on the arithmetically averaged L_{ASmx} values for the pass-by direction which yielded the highest average value for constant high speed pass-bys [30 mph (nominal)] for the 50 foot microphone. Averages of the measured speeds are specified in the last column. Four of the snowcoaches were measured at each of the three site locations and Section 4.3 presents the difference in sound levels due to site bias.

The Alpen Guide – Kitty and Yellowstone Expedition – Hayden snowcoaches consistently had lower sound levels at both speeds, in comparison to the other snowcoaches. The Xanterra 707, Xanterra 709, See Yellowstone Tours #6 and See Yellowstone Tours #9 snowcoaches had high sound levels at both speeds.

Table 14. L_{ASmx} for Loudest Side of Vehicle at Low Speed, dBA

Vehicle	Entrance	Vehicle Side*	50 Foot Average L_{ASmx} (dBA)	100 Foot Average L_{ASmx} (dBA)	Average Speed of Runs (mph)
See Yellowstone Tours #9	West	Right	76	68	15
73 DBA BAT Limit with 2dB Tolerance					
Xanterra 537	North	Right**	71	65	10.7***
See Yellowstone Tours #6	West	Right	69	63	16
Xanterra 709	South	Left**	68	62	15
Xanterra 707*****	North	Right	68	62	15
Buffalo Bus Touring #3	West	Right**	67	59	15
Buffalo Bus Touring T2	West	Left	67	59	16
Xanterra 713*****	South	Right	66	60	16
Yellowstone Snowcoach – SNOVAN5	West	Left	66	59	15
Xanterra 713*****	West	Right	66	60	15
Yellowstone Snowcoach – SNOVAN5	North	Left	66	60	16
Buffalo Bus Touring #4	West	Left	65	59	16
Xanterra 430	North	Left	65	57	17
Yellowstone Expedition – Eleanor	West	Left	65	59	15
Xanterra 713*****	North	Right	65	59	15
Xanterra 710*****	West	Left	64	58	15
Yellowstone Expedition – Hayden	West	Right	64	57	16
Yellowstone Snowcoach – SNOVAN4	North	Left	63	57	15
Alpen Guide – Kitty*****	West	Left	61	55	16
Yellowstone Expedition – Hayden	North	Left	60	55	15
Yellowstone Snowcoach – SNOVAN5	South	Right	59	48	15
Alpen Guide – Kitty*****	North	Left**	59	53	15
Yellowstone Expedition – Hayden	South	Left	59	50	15
Alpen Guide – Kitty*****	South	Right**	58	47	15
See Yellowstone Tours #4	West	N/A****	N/A****	N/A****	N/A****

* "Left/Right" indicates left/right side of vehicle from the driver's perspective.
** Indicates that data was only available for one side of the vehicle.
*** The Xanterra 537 has a top speed of approximately 16 mph. This speed was captured for the development of a noise-distance relationship (see Section 5.1).
**** See Yellowstone Tours #4 did not have a low speed series due to the accumulation of snow on the windscreens.
***** These vehicles have been retrofitted for a reduced sound level.

Table 15. L_{ASmx} for Loudest Side of Vehicle at High Speed, dBA

Vehicle	Entrance	Vehicle Side*	50 Foot Average L_{ASmx} (dBA)	100 Foot Average L_{ASmx} (dBA)	Average Speed of Runs (mph)
Xanterra 707****	North	Left	80	75	29
Xanterra 709	South	Left	80	74	28
Yellowstone Snowcoach – SNOVAN5	West	Left	78	71	36
See Yellowstone Tours #9	West	Left	78	71	36
See Yellowstone Tours #6	West	Right	78	71	35
Xanterra 713****	West	Right	77	73	32
Yellowstone Snowcoach – SNOVAN5	North	Right	77	72	34
Xanterra 537	North	Left	77	71	16***
Xanterra 710****	West	Right	76	71	32
Yellowstone Snowcoach – SNOVAN4	North	Right	76	70	33
Xanterra 713****	North	Right	76	71	26
73 DBA BAT Limit with 2dB Tolerance					
Xanterra 713****	South	Left	75	69	23
Buffalo Bus Touring #3	West	Left	75	67	27
Buffalo Bus Touring T2	West	Left	72	64	24
See Yellowstone Tours #4	West	Right	72	64	33
Buffalo Bus Touring #4	West	Left	72	67	24
Xanterra 430	North	Left	71	65	23
Yellowstone Expedition – Hayden	West	Left	70	63	30
Yellowstone Expedition – Eleanor	West	Left	69	62	25
Yellowstone Snowcoach – SNOVAN5	South	Right	68	58	25
Alpen Guide – Kitty****	West	Right	67	60	32
Alpen Guide – Kitty****	North	Right	67	63	29
Alpen Guide – Kitty****	South	Right**	66	56	32
Yellowstone Expedition – Hayden	North	Left	64	59	20
Yellowstone Expedition – Hayden	South	Left	64	55	20

* "Left/Right" indicates left/right side of vehicle from the driver's perspective.
** Indicates that data was only available for one side of the vehicle.
*** The Xanterra 537 has a top speed of approximately 16 mph. This speed was captured for the development of a noise-distance sound level curve (see Section 5.1).
**** These vehicles have been retrofitted for a reduced sound level.
Note: SAE J1161 allows for a two-decibel exceedance, snowcoaches that pass the BAT within this tolerance are shaded yellow.

The tables show the only snowcoach to fail the 73 dBA BAT requirement for the 15 mph pass-by is the See Yellowstone Tours #9 snowcoach. 12 of the snowcoaches meet the 73 dBA BAT requirements for the high speed pass-by events. Two more snowcoaches meet the requirement, highlighted in yellow, with the 2 dB exception. This BAT requirement may be satisfied if a speed restriction is placed on some of the snowcoaches (e.g., Yellowstone Snowcoach – SNOVAN5). Examining the high speed pass-bys for SNOVAN5 it can be seen that it produces a sound level of about 68 dBA at 25 mph and the sound level increases to 77 dBA at 34 mph.

Table 16 rank orders the twenty five snowcoaches measured from highest to lowest sound level based on the L_{Aeq} values for the idle orientation for the 50 foot microphone. In this case, instead of averaging pass-by measurements, a one-minute period of $L_{Aeq,1sec}$ values were averaged. The idle sound levels of several snowcoaches were within 10 dBA of the ambient, indicating that at least some level of contamination, the maximum ambient sound level for the day is listed next to the idle sound level. Contamination may cause these sound levels to overestimate the true idle sound levels of these vehicles, so these sound levels should be considered a conservative estimate of the idle sound levels. Idle sound levels within 10 dB of the ambient are marked with an asterisk (*) in Table 16. Whereas sound levels during pass-by measurements are due to a combination of engine / exhaust, drive train, and ski / track noise, idle sound levels are predominantly due to engine / exhaust noise alone.

Table 16. Idle L_{Aeq} for Loudest Side of Vehicle at Idle (0 mph), dBA

Vehicle	Entrance	Vehicle Side*	50 Foot Average L_{ASmx} (dBA)	100 Foot Average L_{ASmx} (dBA)	Daily Ambient Sound Level Range (dBA)	
					Minimum	Maximum
See Yellowstone Tours #9	West	Left	57	54	26	33
Xanterra 709	South	Left	49	42	22	24
Xanterra 537	North	Left	47	41	28	34
Xanterra 707***	North	Left	47	45	28	34
Xanterra 713***	West	Left	46	41	22	24
Xanterra 713***	North	Left	45**	41	32	42
Xanterra 710***	West	Right	45	39	23	35
Buffalo Bus Touring #4	West	Right	44**	39	23	35
Xanterra 713***	South	Left	43	37	22	24
See Yellowstone Tours #4	West	Right	43**	37	26	33
Yellowstone Expedition – Hayden	North	Left	41**	37	32	42
Yellowstone Snowcoach – SNOVAN5	North	Right	41**	39	28	34
Alpen Guide – Kitty***	North	Right	41**	39	32	42
Yellowstone Snowcoach – SNOVAN4	North	Left	41**	38	32	42
See Yellowstone Tours #6	West	Right	41**	33	23	35
Yellowstone Expedition – Hayden	West	Left	40	35	22	24
Buffalo Bus Touring T2	West	Right	40**	33	23	35
Yellowstone Snowcoach – SNOVAN5	West	Left	40	34	22	24
Alpen Guide – Kitty***	South	Right	39	33	22	24
Yellowstone Expedition – Hayden	South	Left	39	31	22	24
Yellowstone Snowcoach – SNOVAN5	South	Right	38	31	22	24
Alpen Guide – Kitty***	West	Right	38	33	22	24
Yellowstone Expedition – Eleanor	West	Right	37**	32	23	35
Buffalo Bus Touring #3	West	Both	37**	33	26	33
Xanterra 430	North	Right	37**	32	28	34

* "Left/Right" indicates left/right side of vehicle from the driver's perspective.
** Idle sound levels within 10 dB of the ambient sound level
*** These vehicles have been retrofitted for a reduced a sound level.

4.2.4 Sound Level Versus Speed

The pass-by sound levels for low speeds and high speeds were combined with idle sound levels to generate sound level versus speed curves as shown in Figure 39. This figure presents the data from Table 14 though Table 16 graphically for easier comparison as a function of speed. It should be noted that evaluating the curves, based on idle, low and high speeds, is an imperfect method. Measuring more points would create a more realistic representation of snowcoach noise characteristics, but is outside the scope of this project. Previous vehicle studies[10] have shown that an increase in speed versus sound level should roughly follow an increasing monotonic curve. This relationship would allow us to qualitatively describe gross site differences.

The Alpen Guide – Kitty and Yellowstone Expedition – Hayden consistently have the lowest overall noise profile over the speed range tested while the See Yellowstone Tours #9 and Xanterra 537 have the highest. At the low speed (15 mph) it can be seen that most of the snowcoaches are clustered around the 65 dBA sound level. A high speed comparison is more difficult to make as the snowcoaches traveled between 20 – 36 mph, however the slope of the curves are similar for most snowcoaches in this speed range.

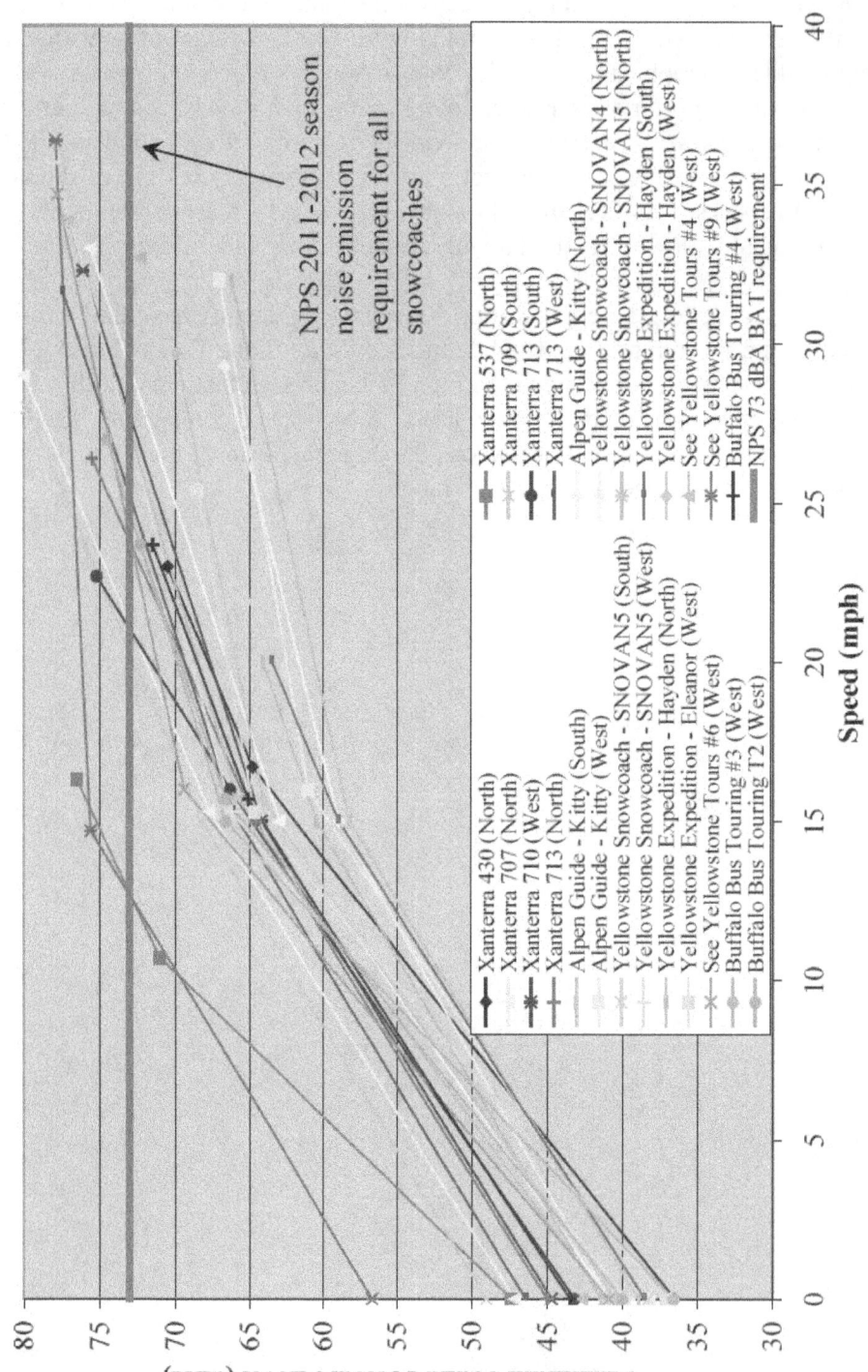

Figure 39. Maximum Sound Level vs. Speed* for All Snowcoaches

* Theoretically the sound level curves should asymptotically approach a constant sound level as the speed approaches 0 mph. However, measurements at more speeds would be required in order to capture the shape of this curve. For modeling purposes in the parks, three points are considered sufficient.

4.2.5 Sound Exposure Level (SEL), dBA

The overall sound exposure level (SEL) due to a snowcoach pass-by gives another indication of the noise performance of the snowcoach. SEL can be estimated from the stored LD824 data by using the following equation:

$$SEL = 10\log_{10}\left(\Delta t \sum_{i=1}^{N} 10^{L_{Aeq\Delta t}(i)/10}\right)$$

Where Δt is the time period for each record, $L_{Aeq\Delta t}(i)$ is the ith $L_{Aeq\Delta t}$ measured, i is the record increment that includes the minimum number of measurements that provide a 10 dB drop from the peak $L_{Aeq\Delta t}$ for both the approach and departure during the pass-by. Because the SEL is an integration of the energy over the exposure period, a vehicle with a slow increase / decrease can have a greater SEL than a vehicle with the highest L_{ASmx}.

The SEL values reported in Table 17 and Table 18 are energy averages of all pass-by measurements that were used in the calculation of L_{ASmx} for Table 14 and Table 15. SEL was not computed for idle events as it has little relevance due to the steady-state nature of the idle data. SEL data is presented as a reference and is not used in INM modeling.

Table 17. Average SEL for each Event at Low Speed, dBA

Vehicle	Entrance	Vehicle Side*	50 Foot Average SEL (dBA)	100 Foot Average SEL (dBA)	Average Speed of Runs (mph)
See Yellowstone Tours #9	West	Right	81	76	15
Xanterra 537	North	Right	81	78	11
Xanterra 430	North	Right	77	66	16
Xanterra 707****	North	Left	76	72	15
See Yellowstone Tours #6	West	Right	76	71	16
Xanterra 709	South	Left	75	71	15
Buffalo Bus Touring #3	West	Left	74	69	15
Buffalo Bus Touring T2	West	Left	74	68	16
Xanterra 713****	West	Right	73	70	15
Yellowstone Snowcoach – SNOVAN5	West	Right	73	68	15
Xanterra 713****	South	Right	73	67	16
Xanterra 713****	North	Right	73	70	15
Yellowstone Snowcoach – SNOVAN5	North	Left	73	69	16
Buffalo Bus Touring #4	West	Left	72	69	16
Xanterra 710****	West	Left	72	69	15
Yellowstone Expedition – Eleanor	West	Left	71	68	15
Yellowstone Expedition – Hayden	West	Right	71	66	16
Yellowstone Snowcoach – SNOVAN4	North	Left	70	66	15
Alpen Guide – Kitty****	West	Left	68	65	16
Yellowstone Expedition – Hayden	North	Left	68	64	15
Alpen Guide – Kitty****	North	Left	66	63	15
Yellowstone Snowcoach – SNOVAN5	South	Right	66	57	15
Yellowstone Expedition – Hayden	South	Left	66	58	15
Alpen Guide – Kitty****	South	Right	64	55	15
See Yellowstone Tours #4	West	N/A***	N/A***	N/A***	N/A***

* "Left/Right" indicates left/right side of vehicle from the driver's perspective.
** Indicates that data was only available for one side of the vehicle.
*** See Yellowstone Tours #4 did not have a low speed series due to inclimate weather.
**** These vehicles have been retrofitted for a reduced sound level.

Table 18. Average SEL for each Event at High Speed, dBA

Vehicle	Entrance	Vehicle Side*	50 Foot Average SEL (dBA)	100 Foot Average SEL (dBA)	Average Speed of Runs (mph)
Xanterra 707***	North	Left	86	82	29
Xanterra 537	North	Left	85	82	16
Xanterra 709	South	Left	84	79	28
Xanterra 713***	West	Right	83	80	32
Yellowstone Snowcoach – SNOVAN5	West	Left	82	77	36
See Yellowstone Tours #6	West	Right	82	77	35
See Yellowstone Tours #9	West	Left	82	76	36
Yellowstone Snowcoach – SNOVAN5	North	Right	82	78	34
Xanterra 710***	West	Left	82	78	32
Xanterra 713***	South	Left	82	76	23
Xanterra 713***	North	Right	82	78	26
Yellowstone Snowcoach – SNOVAN4	North	Right	80	76	33
Buffalo Bus Touring #3	West	Left	80	74	27
Buffalo Bus Touring T2	West	Left	78	72	24
See Yellowstone Tours #4	West	Right	77	70	33
Buffalo Bus Touring #4	West	Left	77	74	24
Xanterra 430	North	Left	77	73	23
Yellowstone Expedition – Hayden	West	Left	75	70	30
Yellowstone Expedition – Eleanor	West	Left	74	69	25
Yellowstone Snowcoach – SNOVAN5	South	Right	73	65	25
Alpen Guide – Kitty***	North	Right	72	69	29
Alpen Guide – Kitty***	West	Right	72	67	32
Alpen Guide – Kitty***	South	Right	71	62	32
Yellowstone Expedition – Hayden	North	Left	71	68	20
Yellowstone Expedition – Hayden	South	Left	69	62	20

* "Left/Right" indicates left/right side of vehicle from the driver's perspective.
** Indicates that data was only available for one side of the vehicle.
*** These vehicles have been retrofitted for a reduced sound level.

4.2.6 Sound Level Spectra

For each record in the sound level time histories, there is an associated one-third octave band spectrum ranging from 12.5 Hz to 20 kHz. The spectrum associated with the L_{ASmx}* during one pass-by of the Xanterra 710 Bombardier measured at high and low speeds on January 21st is shown as an example in Figure 40. Here, the general pattern of decreasing sound level with increasing one-third octave band center frequency can be seen at both speeds. These spectra contain tones which are associated with engine or drive train harmonics and can be expected to increase in frequency with increasing engine speed. The effect of increasing engine speed on tonality can be seen in Figure 41, where a tone can be seen in the 80 and 160 Hz one-third octave bands at low speed. At the higher speed (increased engine rpm), the tones shift to the 125 and

---

* The $L_{Aeq\Delta t}$ is most appropriate for examining time histories, but the L_{ASmx} gives a better indication of the maximum level during a Pass-by.

250 Hz one-third octave bands, respectively. Although the general trend is for the spectral level to decrease at high frequencies an increase can be seen at the high frequencies. It was verified that this was not due to the ambient.

Unweighted, one-third octave band spectra associated with the pass-by time histories in Appendix E are presented in Appendix F.

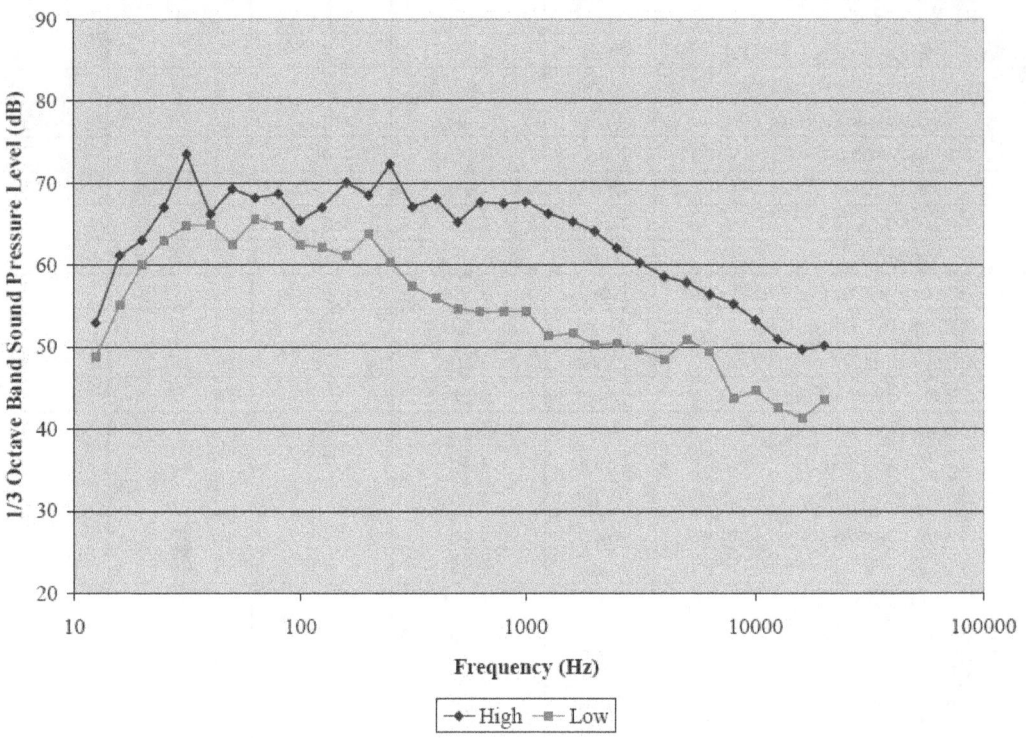

Figure 40. Xanterra 710, West Entrance (Jan 21st) Spectra for Low Speed and High Speed

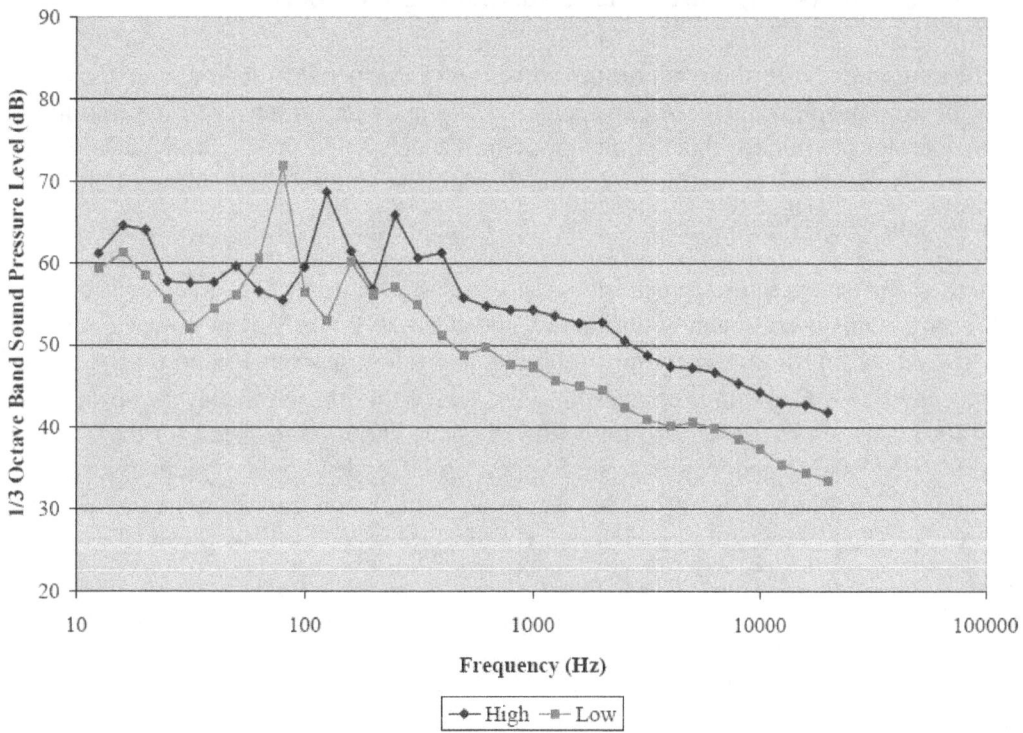

Figure 41. Yellowstone Snowcoach – SNOVAN5, (Jan 14) Spectra for Low Speed and High Speed

4.3 Evaluation of Measurement Site Bias

The NPS wished to determine if there is a difference in the snowcoach sound level that could be attributed to measurement site bias. Four snowcoaches, the Xanterra 713, Yellowstone Snowcoach – SNOVAN5, Alpen Guide – Kitty, and Yellowstone Expedition –Hayden, were tested at the three measurement sites. The same exact vehicles were used at the three sites. This was done to eliminate variability between vehicles of the same model. There are also several other variables that should be noted when comparing results:
- The driver could have affected the test as the drivers were not the same for each vehicle in all cases and;
- Meteorological conditions (see 4.1);
- Track conditions, which directly affected the obtainable speed of the snowcoaches and;
- Variable snow accumulation on the moving parts for the tracks/wheels/track mechanism.

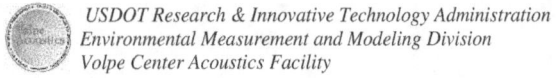
USDOT Research & Innovative Technology Administration
Environmental Measurement and Modeling Division
Volpe Center Acoustics Facility

April 2010

4.3.1 Comparison of 4 Foot Microphone Maximum Sound Levels

Figure 42 though Figure 45 show maximum sound level vs. speed[*] plots for each snowcoach. Note that for low speed events the Alpen Guide – Kitty and the Xanterra 713 have similar levels at all three locations, while for the Yellowstone Snowcoach – SNOVAN5 and Yellowstone Expedition – Hayden the west entrance site was the loudest and the south entrance site was about 7 and 5 dBA quieter, respectively.

For the high speed events the Alpen Guide – Kitty and the Xanterra 713 have very similar levels for all three sites; this is consistent with the low speed events. The Yellowstone Snowcoach – SNOVAN5 and Yellowstone Expedition – Hayden haves levels about 10 and 6 dBA lower, respectively, at the south entrance compared to the west entrance. Note that the speeds at the south entrance were approximately 10 mph slower[†]. The variation in speed for the Yellowstone Snowcoach – SNOVAN5 and Yellowstone Expedition – Hayden could explain at least some of the difference in levels. The slower speeds were due to track conditions on the day of testing, as the track was used it became softer and the maximum safe speed of the snowcoaches was limited.

The idle measurement shows a clear trend that the south entrance was the quietest, by approximately 2 dBA, the exception being the Alpen Guide – Kitty, where the south and west entrances were essentially the same (38 dBA). Vehicle speed is not a contributing factor for idle measurements. This suggests that the snow berm at the south entrance was a contributing factor to the difference in level due to low sound sources being shielded from the microphones, see Section 4.3.2 for further explanation on the snow berm. Other factors such as snow conditions or snow accumulation on the track mechanism, etc., cannot be rejected without more extensive testing.

Additional data pertaining to the site bias issue can be found in Appendix G. This includes a comparison of L_{ASmx} values and a time history plot for each vehicle for each event type.

---

[*] Theoretically the sound level curves should asymptotically approach a constant sound level as the speed approaches 0 mph. However, measurements at more speeds would be required in order to capture the shape of this curve. For modeling purposes in the parks, three points are considered sufficient.
[†] The high speed tests for the Yellowstone Expedition – Hayden at the north and south entrances had similar sound levels and speeds.

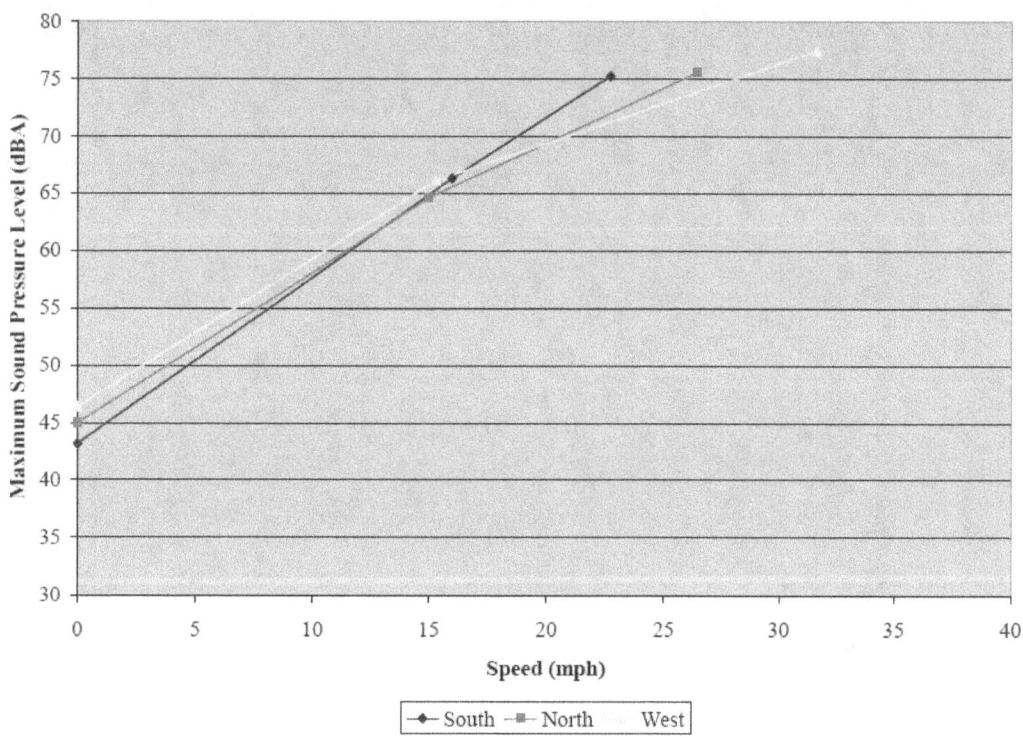

Figure 42. Maximum Sound Level vs. Speed for Xanterra 713

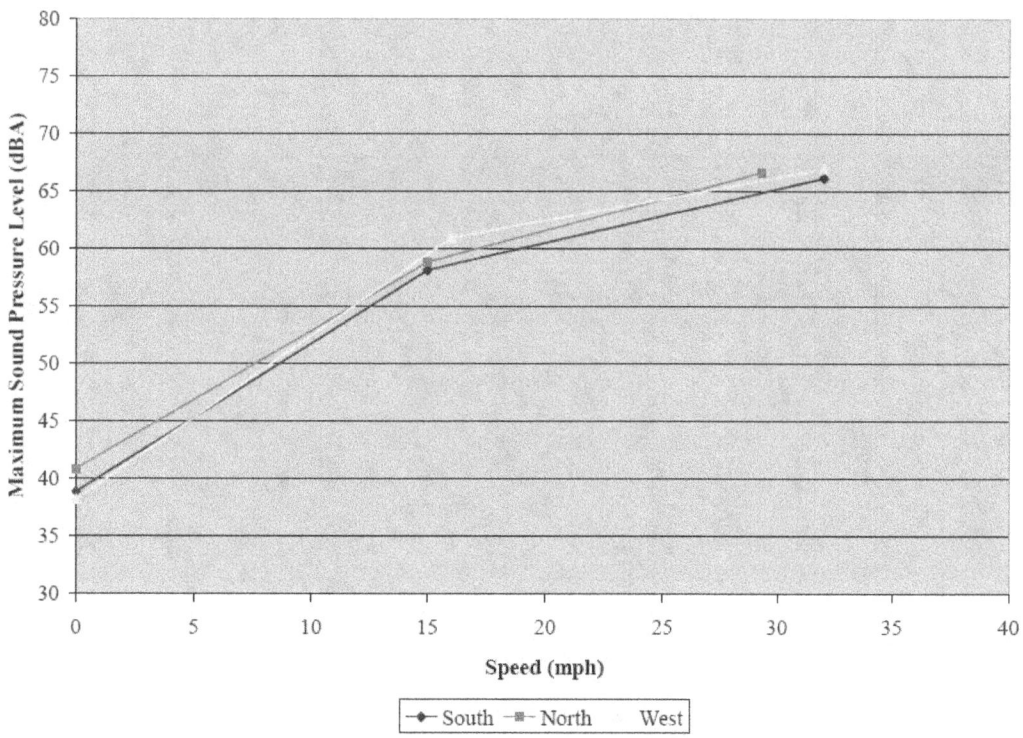

Figure 43. Maximum Sound Level vs. Speed for Alpen Guide – Kitty

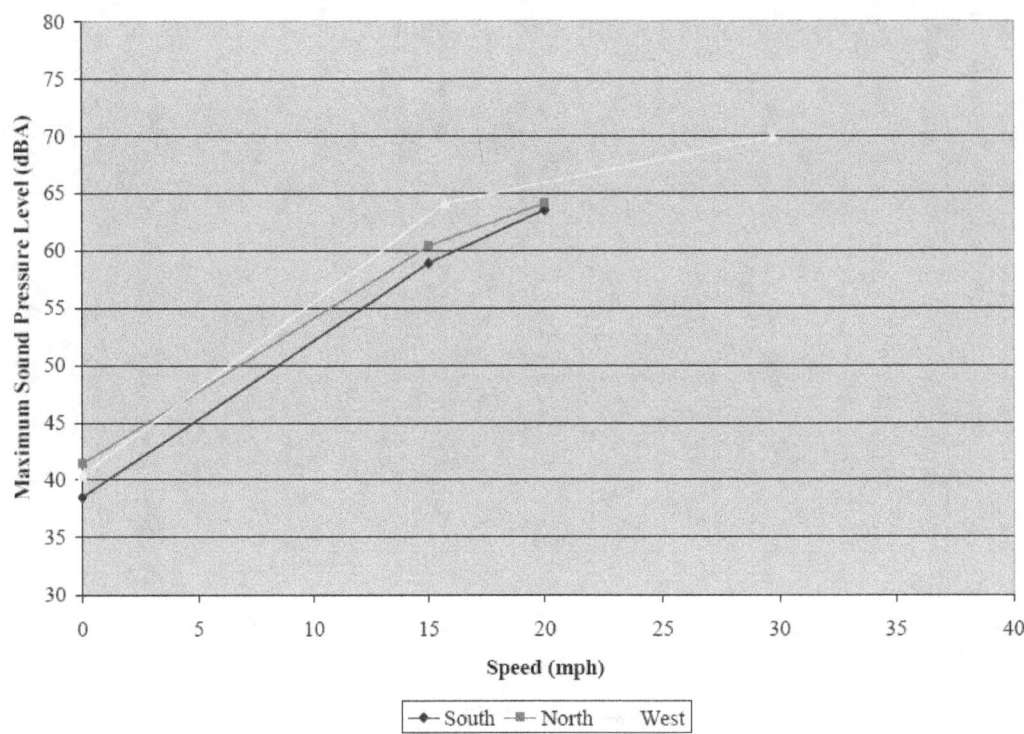

Figure 44. Maximum Sound Level vs. Speed for Yellowstone Expedition – Hayden

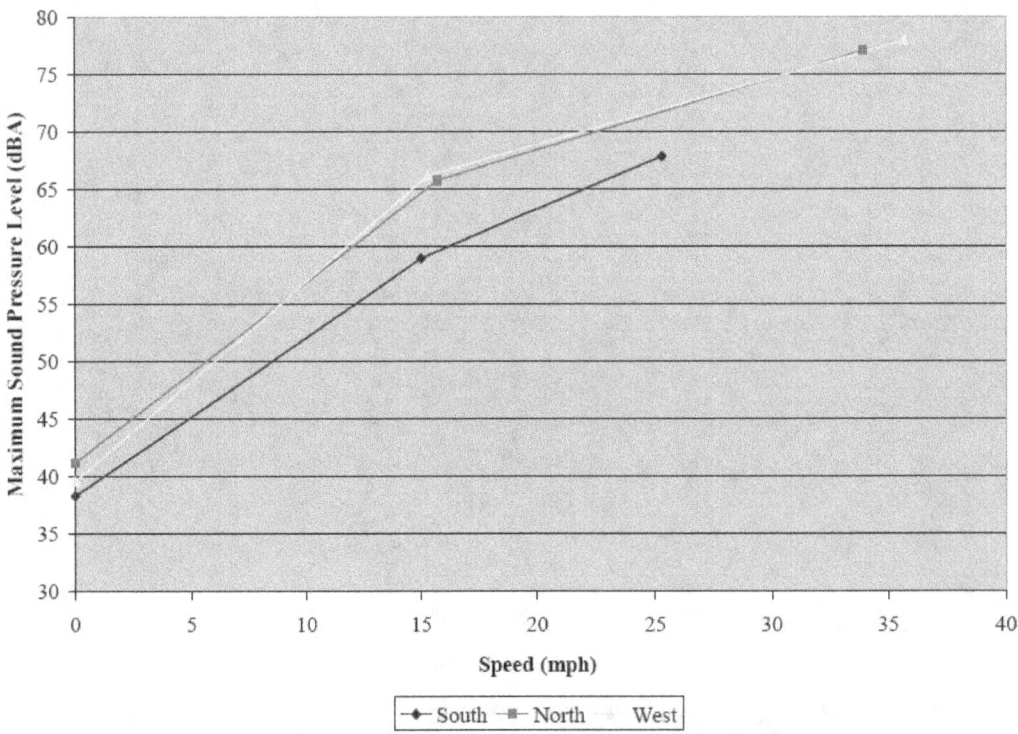

Figure 45. Maximum Sound Level vs. Speed for Yellowstone Snowcoach – SNOVAN5

4.3.2 Comparison of 4 Foot and 15 Foot Microphone Sound Levels at 50 Feet

In this section, data from the 4-foot and 15-foot high microphones are compared. By comparing the results between the south entrance (with snow berm) with those from the north and west entrances (without snow berm), a better understanding of the effects of the snow berm can be developed. Table 19 shows a summary by site and vehicle of the average difference of the maximum record between the 4 and 15 foot microphones for both high and low speed.

Table 19. Average Maximum Sound Level Difference Between 4 and 15 Foot Microphones ($L_{ASmx,4foot} - L_{ASmx,15foot}$) Low Speed and High Speed

Vehicle	Average Difference $L_{ASmx4foot} - L_{ASmx15foot}$ (dBA)		
	South	North	West
Alpen Guide – Kitty	-4.6	-2.0	-2.8
Yellowstone Snowcoach – SNOVAN5	-4.3	N/A*	-2.6
Yellowstone Expedition – Hayden	-3.0	-1.8	-2.6
Xanterra – 713	-1.6	-0.7	-1.0

* Due to equipment issues, no 15 foot microphone data were collected on 1/16/2009

The average difference between the two microphones is about 1.8 dBA greater at the south entrance compared to the north and about 1.1 dB greater at the south entrance compared to the west entrance. This seems to indicate that the snow berm might have introduced a small, but measurable, amount of sound level attenuation. Snow berms are present around Yellowstone National Park during the winter season and can be accounted for during the modeling process. Currently we do not have enough data to properly determine the effect of a snow berm. Future measurements could be conducted to determine the effect of berms for modeling purposes.

It should be noted that the dominant source for each snowcoach was at different locations (heights), where higher sources would be affected less by the snow berm than lower sources. The Xanterra 713 had high stacks emitting sound which were not present on any of the other vehicles. The Alpen Guide – Kitty had mostly engine noise which came from the middle of the snowcoach. The Yellowstone Expedition – Hayden and Yellowstone Snowcoach – SNOVAN5 had a lot of mechanical noise in the track mechanism and would be more affected by the snow berm than the other two coaches.

4.3.3 Barometric Pressure Effects on Measurements

Meteorological conditions must be monitored during measurements as outlined in standard SAE J1161:

> 7.4 Measurements shall be made only when the wind speed is below 19 km/h (12 mile/h) and absolute barometric pressure is between 93 and 103 kPa (27.5 and 30.5 in of mercury).

Wind speed was measured at a rate of 1 sample/second. Acoustic measurements made during wind speeds greater than 12 mph were discarded, in accordance with SAE J1161.

SAE J1161 presents the range of pressure in kilopascal (kPa) and inches of mercury, when converted to millibars the range is 930 – 1030 millibars. Table 20 shows the minimum, maximum, and average recorded barometric pressures for each measurement day. Barometric pressure readings recorded at the three measurement sites indicate that Yellowstone National Park is not normally within the SAE J1161 requirements. The pressure measured in the park during the measurements ranged from 786 to 806 millibars. This lower barometric pressure is due to the elevation above sea level of Yellowstone (approximately 8,000 feet). It can be expected that many places where snowcoaches would be measured would also have conditions below the SAE J1161 standard barometric pressure. The NPS snowmobile BAT sound level requirement specifies a minimum barometric pressure of 793 millibars (23.4 inches Hg) uncorrected as measured at or near the test site. It is recommended that pressure readings be recorded and documented for any future measurements. If the data are subsequently used for modeling, the sound levels can be adjusted to standard conditions. The barometric pressure being below the SAE J1161 standard will not have a significant impact on the measurements as long as there is no sudden change in the pressure. There will be a negligible effect on sound level propagation between the 50 and 100 foot microphone.

Table 20. Summary of Temperature and Pressure for Measurement Periods

Entrance	Date	Barometric Pressure (millibars)		
		Min.	Max.	Avg.
South	1/14/2009	796	797	797
North	1/15/2009	786	787	787
	1/16/2009	787	790	788
West	1/20/2009	803	806	804
	1/21/2009	796	800	798
	1/22/2009	793	794	794

5 Development of Data to Support Future Modeling

Each of the twenty five snowcoaches have been processed and formatted into 4 noise-distance curves: a L_{ASmx} and SEL for both low speed and high speed, and maximum unweighted spectra at 1000 feet at the time of L_{ASmx} for both low speed and high speed. These noise-distance curves were developed for future use by NPS in the modified version of INM 6.2[16].

Noise-Power-Distance relationships are the starting point for INM's propagation algorithms. However, snowcoach speed is a more appropriate parameter than power. Therefore a noise-distance curve for both low speed and high speed was created. The noise-distance relationships are developed by modeling the effects of spherical divergence[17] and frequency dependent atmospheric absorption[18]. Test-day temperature and humidity were input and then adjusted to standard conditions.

Four of the twenty five snowcoaches were tested at all three measurement sites and for these snowcoaches, noise-distance curves and maximum spectra were averaged together. Where speeds did not align multiple curves were created.

5.1 Noise-Distance Curves

The following section shows examples of noise-distance curves. The remaining curves can be found in Appendices G and H for L_{ASmx} and SEL, respectively.

5.1.1 L_{ASmx} Noise-Distance Curves

Exemplar numerical values for the Xanterra 713 L_{ASmx} noise-distance curves can be seen in Table 21. Figure 46 shows these data graphically. L_{ASmx} Noise-distance data and graphs for all events can be seen in Appendix H.

Table 21. Xanterra 713 Maximum Sound Level Noise-Distance Curve Values

Vehicle	Xanterra 713		
Speed (mph)	15	25	32
Distance (Feet)	Maximum Sound Pressure Level (dBA)		
200	53.2	62.9	65.1
400	46.8	56.4	58.6
630	42.4	51.9	54.1
1000	37.8	47.2	49.3
2000	30.7	39.6	41.5
4000	23.0	31.3	32.7
6300	17.6	25.3	26.2
10000	11.8	18.6	18.8
16000	5.6	11.4	10.6
25000	-0.8	4.2	2.2

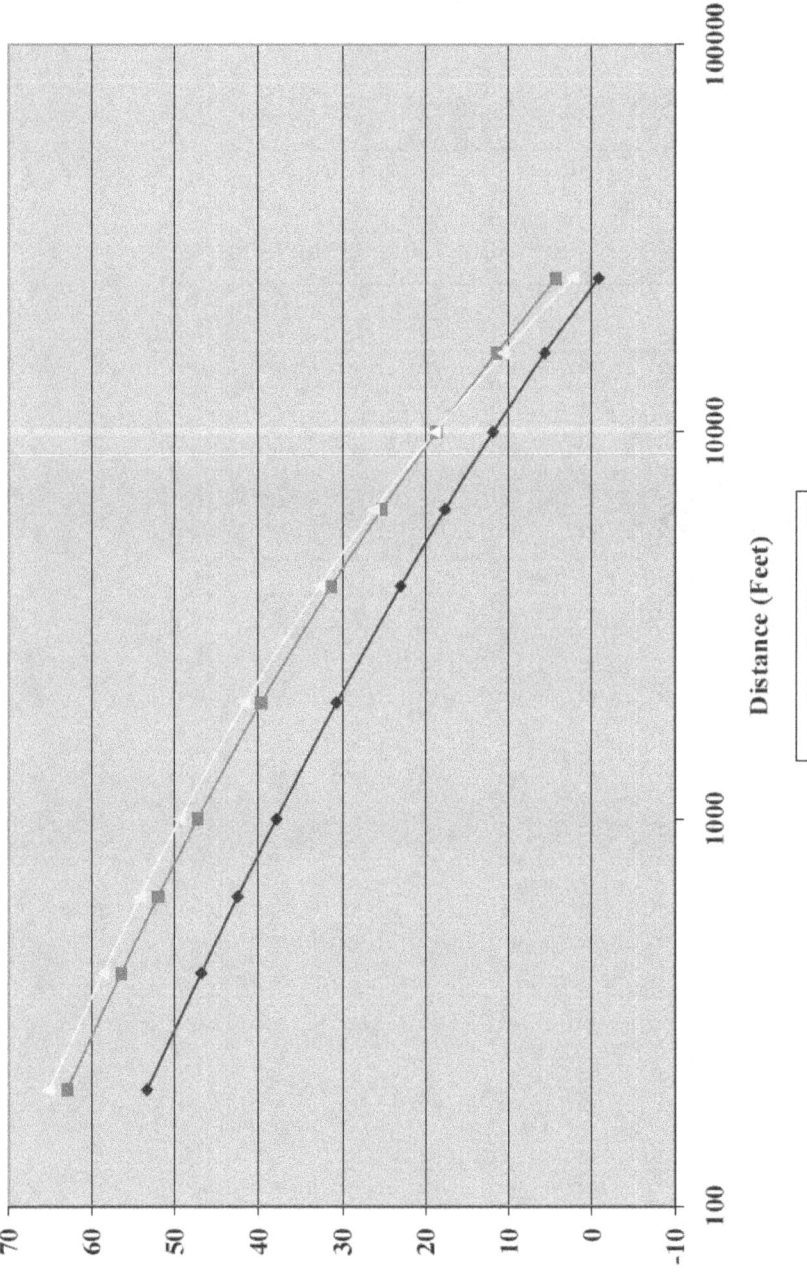

Figure 46. Xanterra 713 Maximum Sound Pressure Level Noise-Distance Curves

5.1.2 SEL Noise-Distance Curves

Exemplar numerical values for the Xanterra 713 SEL noise-distance curves can be seen in Table 22. Figure 47 shows these data graphically. SEL Noise-distance data and graphs for the three measurement sites can be seen in Appendix I.

Table 22. Xanterra 713 SEL Noise-Distance Curve Values

Vehicle	Xanterra 713		
Speed (mph)	15	25	32
Distance (Feet)	Sound Exposure Level (dB)		
200	54.3	64.8	67.5
400	50.1	60.5	63.2
630	47.2	57.5	60.2
1000	44.1	54.3	56.9
2000	39.2	49.0	51.4
4000	33.8	42.9	44.8
6300	29.9	38.4	39.8
10000	25.6	33.2	33.9
16000	20.9	27.6	27.3
25000	15.9	21.8	20.3

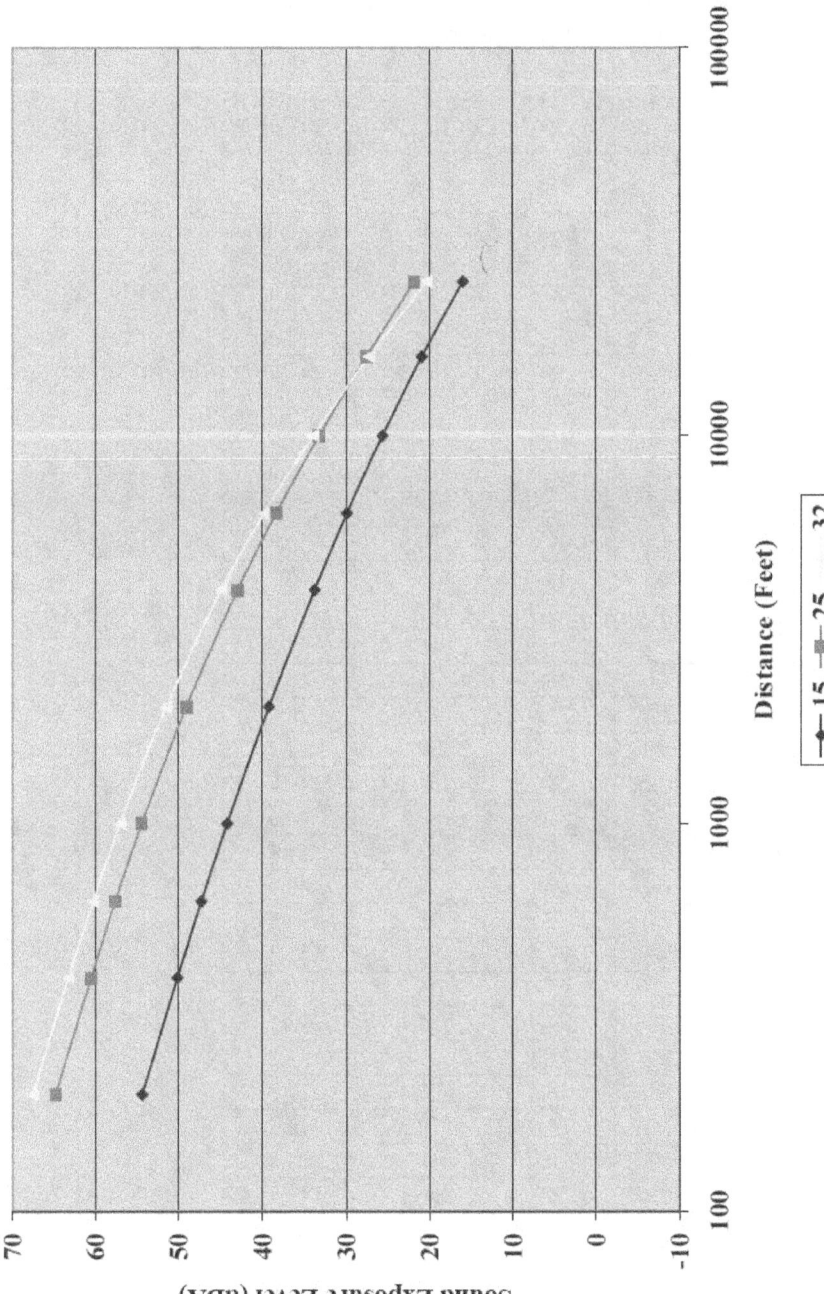

Figure 47. Example Low Speed SEL Noise-Distance Curves

5.2 Spectral Data

Exemplar numerical values for the Xanterra 713 spectra at 1000 feet at the time of L_{ASmx} can be seen in Table 23. Figure 48 shows these data graphically. Spectra at 1000 feet at the time of L_{ASmx} and graphs for all events can be seen in Appendix J.

Table 23. Xanterra 713 Maximum Spectra at 1000 Feet

Vehicle		Xanterra 713		
	Speed (mph)	15	25	32
		Spectra Sound Pressure Level at 1000 feet at time of LASmx (dB)		
Frequency (Hz)	12.5	25.8	32.2	25.1
	16	33.7	33.4	33.5
	20	38.6	38.9	42.2
	25	38.1	45.9	42.2
	31.5	43.9	37.0	48.7
	40	35.4	35.2	39.2
	50	37.6	44.4	37.7
	63	40.6	37.9	40.6
	80	45.7	42.5	36.3
	100	34.7	41.5	39.5
	125	35.5	45.3	37.3
	160	36.2	35.9	46.1
	200	39.9	44.5	38.0
	250	36.2	39.9	42.2
	315	30.6	34.7	39.0
	400	31.6	39.1	40.3
	500	28.9	38.9	44.3
	630	29.1	42.9	40.5
	800	29.2	38.6	40.8
	1000	27.5	37.0	42.4
	1250	25.6	36.8	39.0
	1600	24.4	35.4	38.5
	2000	21.1	31.6	35.2
	2500	18.5	28.4	32.8
	3150	14.9	24.8	30.1
	4000	10.0	18.8	25.8
	5000	9.9	16.1	23.9
	6300	3.3	9.0	17.4
	8000	-6.0	-1.0	10.4
	10000	-15.1	-11.7	0.8

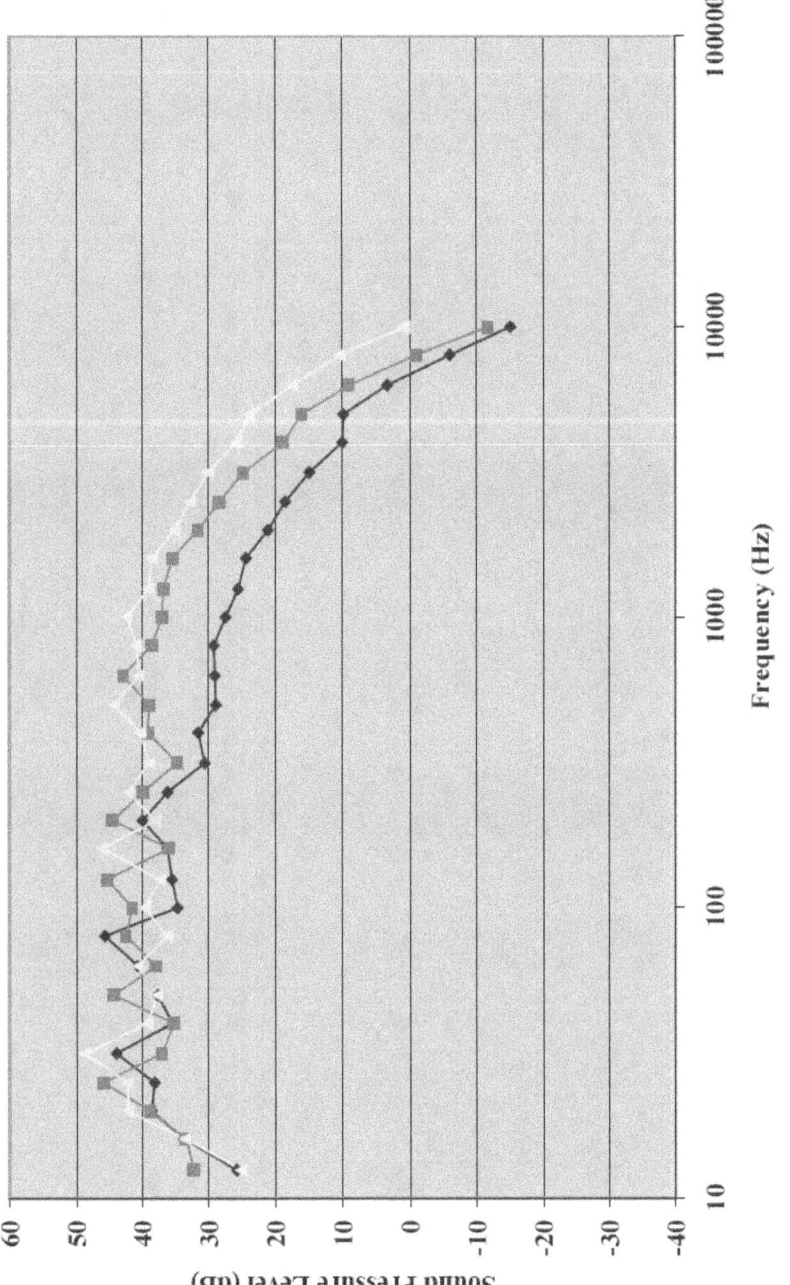

Figure 48. Xanterra 713, 1000 Feet, Maximum Spectra

This page intentionally left blank

6 Summary and Conclusions

Maximum A-weighted sound pressure levels with slow time response were measured for twenty five snowcoaches for low speed (nominally 15 mph) and high speed (nominally 30 mph or full speed) pass-by events. Measurements were made for both sides of each vehicle; the side with the highest average (based on three measurements within 2 dBA) was assigned as the vehicle's pass-by sound level. Idle sound levels were also measured. Examination of the low speed pass-by events indicates that only the See Yellowstone Tours #9 snowcoach exceeded the NPS' 73 dBA BAT requirement. For the high speed pass-by events only 12 coaches meet the NPS' 73 dBA BAT requirement; two additional vehicles meet the requirement with the 2 dB exception. Sound level data indicate that snowcoaches that do not meet the BAT requirement at high speed could adhere to the requirement if a speed restriction for each vehicle were implemented.

Four snowcoaches were measured at three different locations, in order to determine if site location impacted sound levels. Analysis indicated that location does have some impact on sound levels. The factors that have the most impact on sound level are speed, track conditions, snow accumulation on the track mechanism and presence of snow berms/obstructions. Additionally:
- As the speed of the snowcoach increases, the engine must work harder, which contributes to increased sound levels.
- Another factor that has a large impact is the track condition. After the testing of several snowcoaches, the path starts to deteriorate causing the vehicles to either not obtain maximum speed or to work harder to obtain the desired speed. When the path conditions deteriorate the snowcoach must work harder and speed and sound levels have more variance and require more pass-by's to get three good samples.
- Snow accumulation in the track mechanism can attenuate sounds coming from the track mechanism.
- Snow berms or other obstructions may increase ground affects and attenuate sound levels.

A groomer should be utilized to minimize track condition deterioration during sound level testing. It was helpful to have a groomer on hand to smooth out the track and create a flat surface for the snowcoaches. The use of a groomer was most effective at the end of the day when the track had a chance to freeze over night. When it was groomed during the day the snow was still soft and developed ruts after just a few pass-by's. For future measurements, it is recommended that a limited number of coaches be tested per day at any given test site. It is also recommended to groom the site and let it freeze overnight for subsequent measurements.

Barometric pressure readings recorded at the three entrances during testing indicate that Yellowstone National Park is not normally within the SAE J1161 requirements of 930 to 1030 millibars. The pressure measured in the park during the measurements ranged from 786 to 806 millibars. It is recommended that pressure readings be recorded and documented for any future measurements. The NPS snowmobile BAT sound level requirement specifies a minimum barometric pressure of 793 millibars (23.4 inches Hg) uncorrected, which could be used for snowcoaches. If the data is subsequently used for modeling, these sound levels can be adjusted to standard conditions.

Based on the 2008 and 2009 OSV testing at Yellowstone and Grand Teton National Parks, future measurements should adhere to SAE J1161 with the following modifications and considerations:

- Due to the altitude of the parks, the barometric pressure tolerances during measurements should be expanded to include pressures typical of those experienced in the Yellowstone during the winter season[*]. The affects on sound levels due to barometric pressure could be accounted for in a manner similar to the methods described in References 11 and 12.
- Testing should be conducted for three conditions:
    - Idle
    - 15 mph
    - A high speed, to be determined by the park based on local speed limits, e.g., 30 mph, or a typical cruising speed.
- Ambient measurements should be taken regularly (at least hourly) to accurately represent changing conditions throughout the day.
- In order to standardize measurement sites the following criteria need to be followed:
    - Measurements should not be performed at sites with snow berms[†]. The 50 foot microphone should have a clear line of site to any potential noise source on the test snowcoach.
    - A groomer should be utilized to minimize track condition deterioration during testing. A limited number of coaches per day should be tested at any given location depending on the track conditions.
    - Sites should be away from noise sources that could potentially interfere with data collection (e.g. running water).
    - Sites should be free of obstructions (e.g. trees and structures).
- If a vehicle fails to meet BAT requirements at the high speed, consideration should be given to:
    - Restrictions that would still allow the snowcoach to operate in the parks, but at reduced speeds;
    - Modifying the vehicles for a reduce sound level; or
    - Removing the vehicle from the fleet and replacing it with a quieter one.

---

[*] This level could be set at 23.4 Hg (792 millibars) which is the standard in the BAT requirement for snowmobiles.
[†] Snow berms are present around Yellowstone National Park during the winter season and can be accounted for during the modeling process. Currently we do not have enough data to properly determine the effect of a snow berm. Future measurements could be conducted to determine the effect of berms for modeling purposes.

Appendix A: Larson Davis Model 824 Sound Level Meter Settings

This appendix provides the settings used for the Larson Davis Model 824 sound level meter during the measurements documented herein.

One second samples were used for collection. The following metrics were stored with the same time periods as the captured data:

- Average A-weighted sound level (L_{Aeq}),
- Maximum A-weighted slow and fast values (L_{ASmax}, L_{AFmax}),
- Minimum A-weighted slow and fast values (L_{ASmin}, L_{AFmin}).

Table 24. Larson Davis 824 real time analyzer settings

Setting Parameter	Broadband	Narrow Band
Detector	Slow	Slow
Weighting	A	Flat
Bandwidth	N/A	1/3
Period (seconds)	1	1
Resolution (dB)	0.1	0.1

This page intentionally left blank

Appendix B: Sony Model TCD-100 DAT Recorder Settings

This appendix provides the settings used for the Sony Model TCD-100 DAT recorders during the measurements documented herein.

Sampling Frequency	Input	Microphone Attenuation	Input Range
44.1 kHz	Mic	0 dB	-6 dB @ 94 dB, 1 kHz

Sampling rate: 44.1 kHz (CD compatible) mode
Input mode: Always set to "LINE"
Gain control: Always set to "MANUAL"

Tape duration: A 60-m (197-ft) tape will run for 2 hours at normal speed; 4 hours at LP half speed. Although the "LP" mode does not use linear PCM encoding, testing has established that amplitude linearity is good to within +/- 1.5 dB down to 85 dB below full-scale (0 VU).

This page intentionally left blank

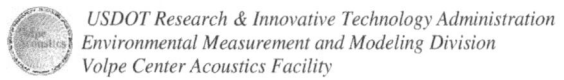
USDOT Research & Innovative Technology Administration
Environmental Measurement and Modeling Division
Volpe Center Acoustics Facility

Appendix C: Measurement Protocol and Logging Procedure

In order to facilitate measurement of OSV's, the following protocol was instituted and found to be effective[*]:

- NPS and Volpe personnel determine what type of measurement will be conducted [15mph, 30mph (full speed), or Idle].
- NPS personnel specify vehicle to be measured.
- Volpe personnel note description of vehicle, check wind speed, and indicate readiness for measurement.
- NPS personnel indicate vehicle is beginning run.
- NPS personnel indicate vehicle speed to the OSV driver to allow the driver to maintain constant speed during pass-by[†].
- Volpe personnel:
    - note the rise and fall of the event sound level on the LD824 as the vehicle travels through the measurement area,
    - note the maximum sound level displayed on the LD824,
    - confirm that the sound level at the start point and stop point were less than the maximum sound level by at least 10 dB, and
    - indicate any potentially contaminating sounds or wind levels greater than 12 mph, identifying the measurement as bad if appropriate.
- NPS personnel indicate if the vehicle strayed from a straight path along the travel path or if speed varied more than 2 mph from target speed, identifying the run as bad if appropriate.
- For the case of constant speed runs, NPS personnel indicate speed and Volpe personnel record the speed.
- Volpe personnel monitor the number of runs by each vehicle in each direction and announce when sufficient measurements are made for the specified vehicle under the specified operating condition.

---

[*] Communication between NPS and Volpe personnel was largely conducted by using radio transceivers. Radio silence was observed during measurement intervals.
[†] Vehicle speed was monitored by means of a hand held GPS device.

Acoustics System Log

Volpe Center
Acoustics Facility

Date:		Acoustics System:		Procedural Checklist: Adjust Calibration	Page ___ of ___
Site Name:		Aircraft:		*Record/Collect:* 1) Cal Tone 4) Cal Tone 2	
Site ID:		Personnel:		2) Mic Simulator 5) Data/Events	
				3) Pink Noise 6) End Cal Tone	

Event	Time (hh:mm:ss)		DAT ID	LD824 Gain Range	LD824 Levels		DAT Input Level (dB)	dB Up/Down	L_{max}	Comments
	Start	End			SLM	RTA @ 1K	Ch 1			

USDOT Research & Innovative Technology Administration
Environmental Measurement and Modeling Division
Volpe Center Acoustics Facility

April 2010

Appendix D: Sound Level Time Histories

Sample time histories for each vehicle at idle, low, and high speed are provided in this appendix. One sample is provided for each test condition for each vehicle (provided the data are available). South entrance sound level time histories are provided in Figure 49 to Figure 53 for each snowcoach. North entrance sound level time histories are provided in Figure 54 to Figure 61 for each snowcoach with available high speed data. West entrance sound level time histories are provided in Figure 62 to Figure 73 for each snowcoach.

South Entrance Sound Level Time Histories

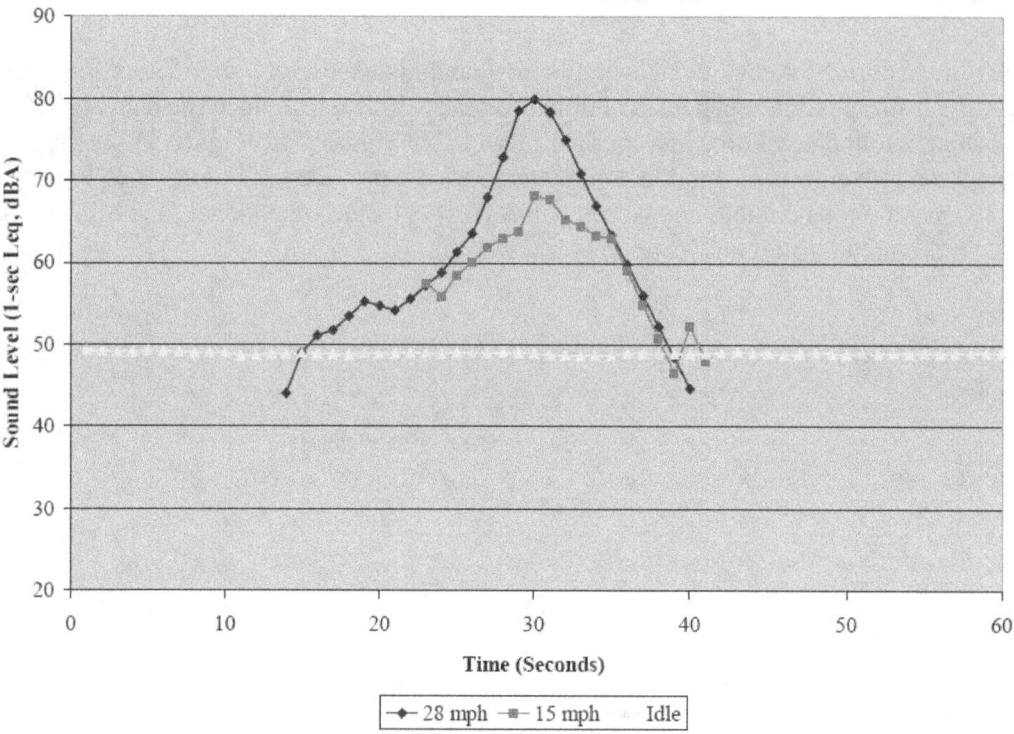

Figure 49. Xanterra 709, South Entrance, (Jan 14th) Time History Plots

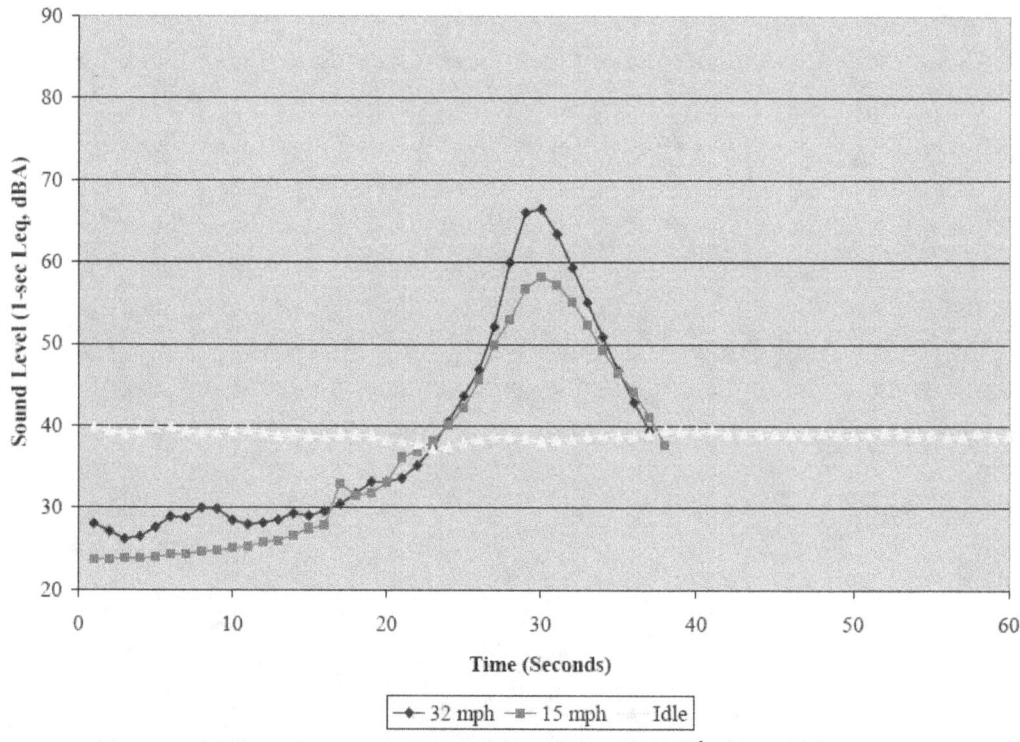

Figure 50. Alpen Guide – Kitty, South Entrance, (Jan 14th) Time History Plots

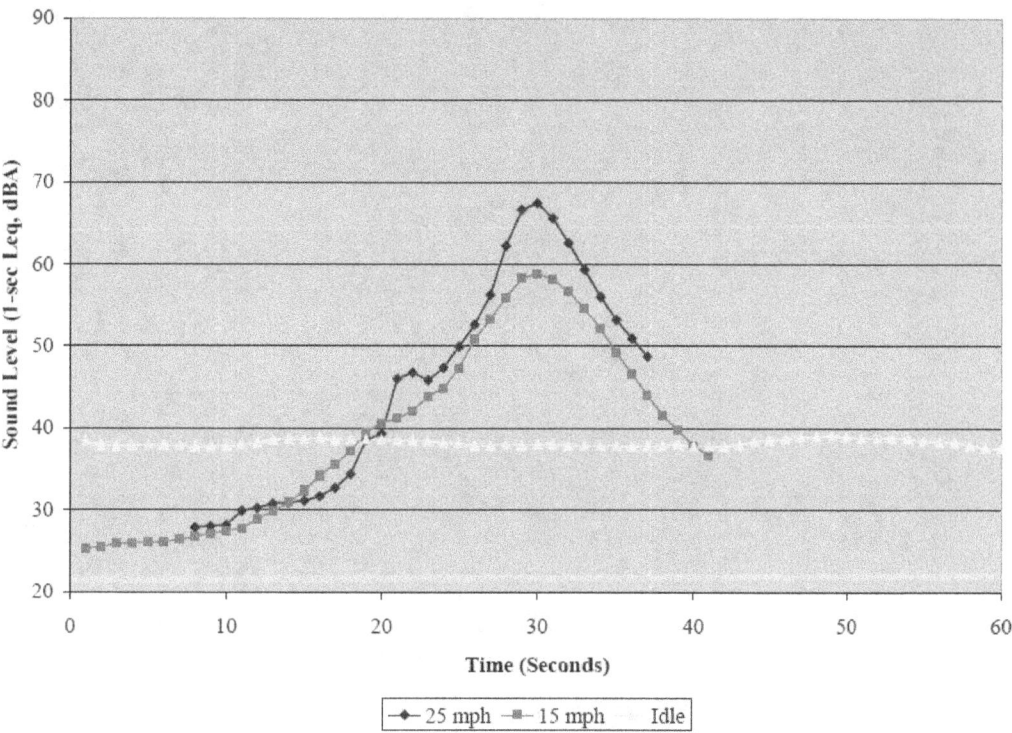

Figure 51. Yellowstone Snowcoach – SNOVAN5, South Entrance, (Jan 14th) Time History Plots

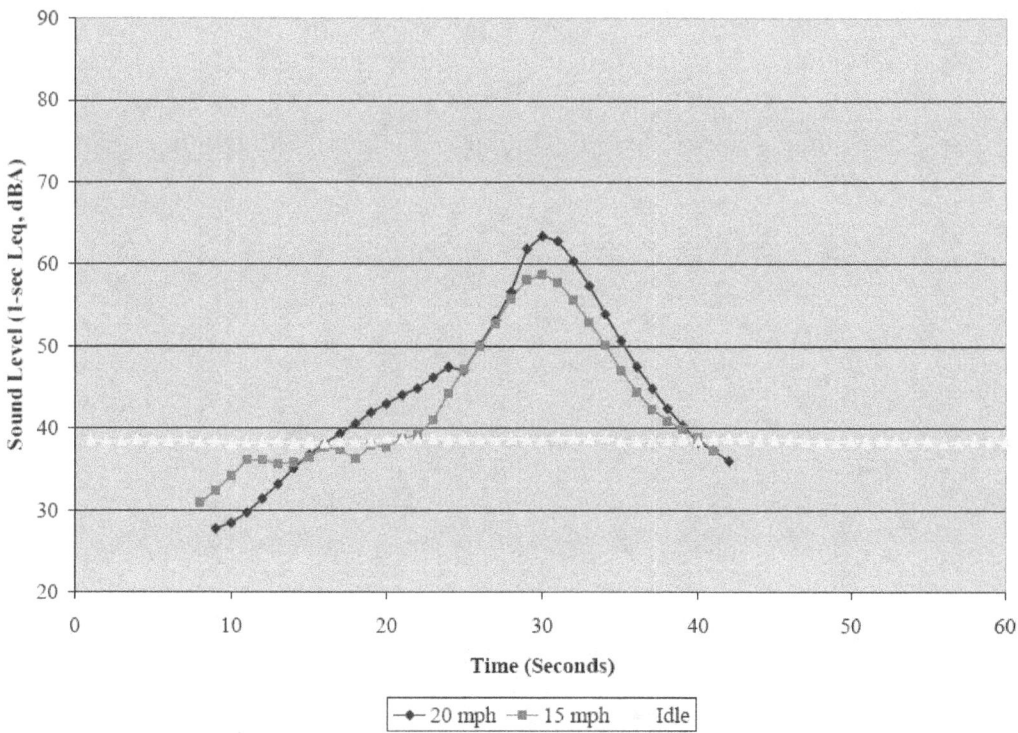

Figure 52. Yellowstone Expedition – Hayden, South Entrance, (Jan 14th) Time History Plots

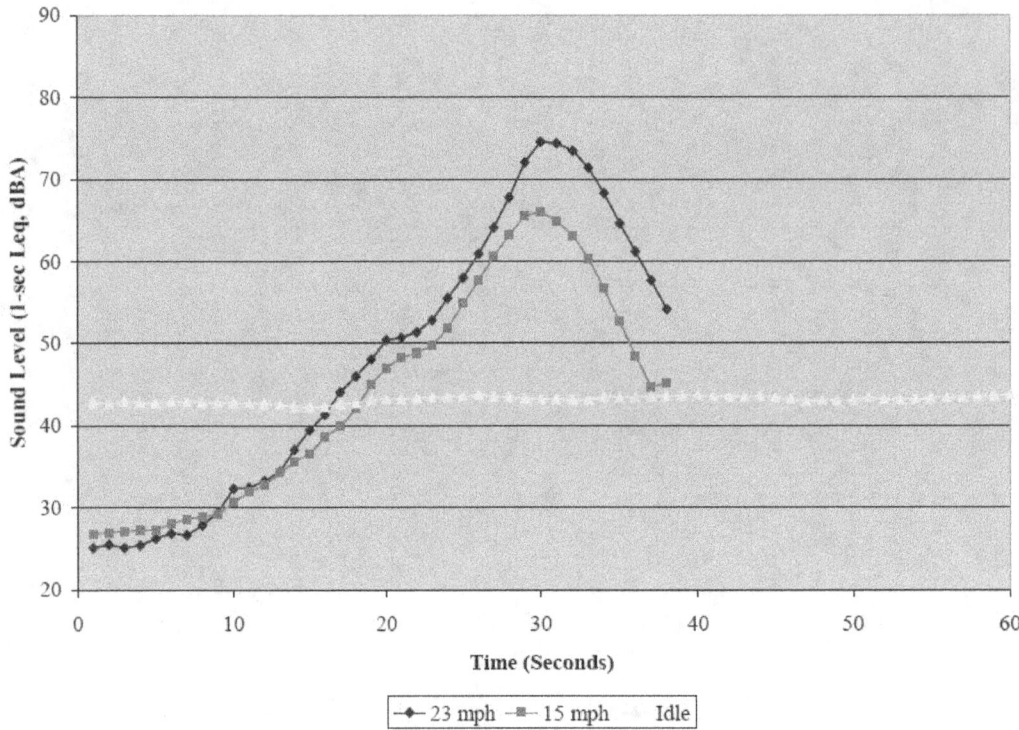

Figure 53. Xanterra 713, South Entrance, (Jan 14th) Time History Plots

North Entrance Sound Level Time Histories

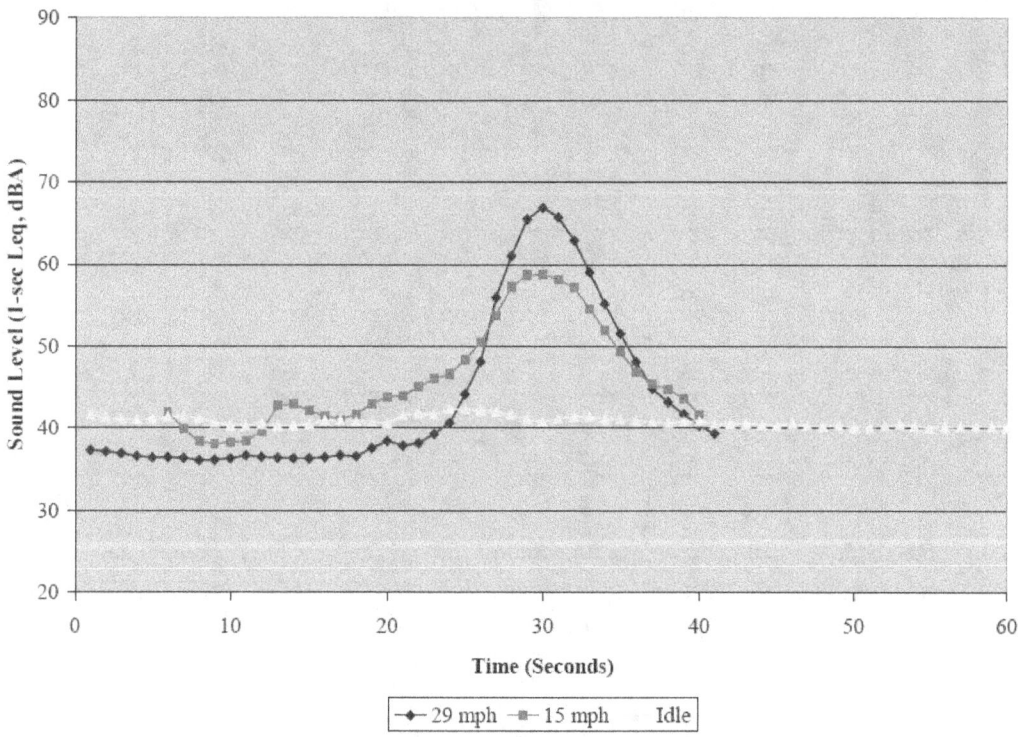

Figure 54. Alpen Guide – Kitty, North Entrance, (Jan 15th) Time History Plots

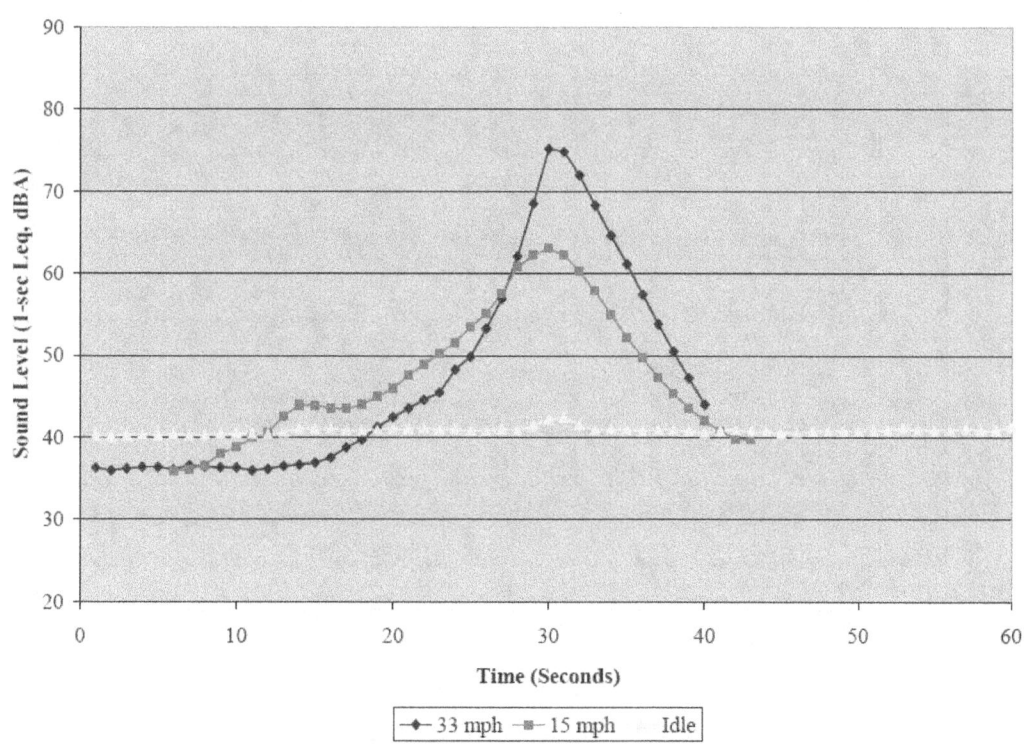

Figure 55. Yellowstone Snowcoach – SNOVAN4, North Entrance, (Jan 15th) Time History Plots

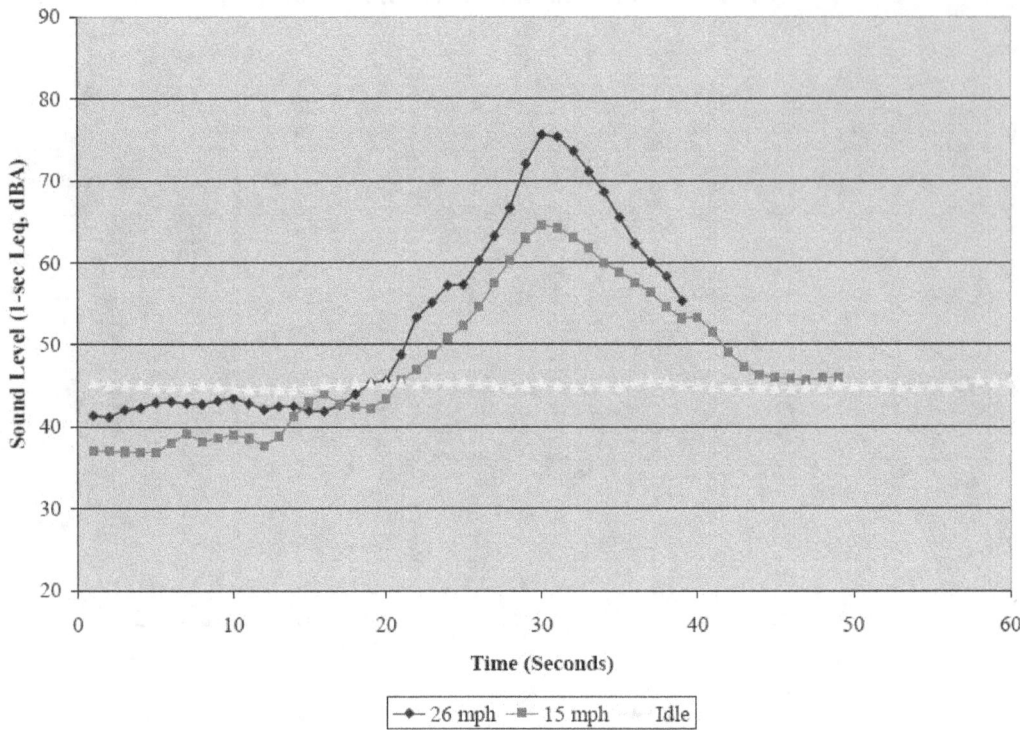

Figure 56. Xanterra 713, North Entrance, (Jan 15th) Time History Plots

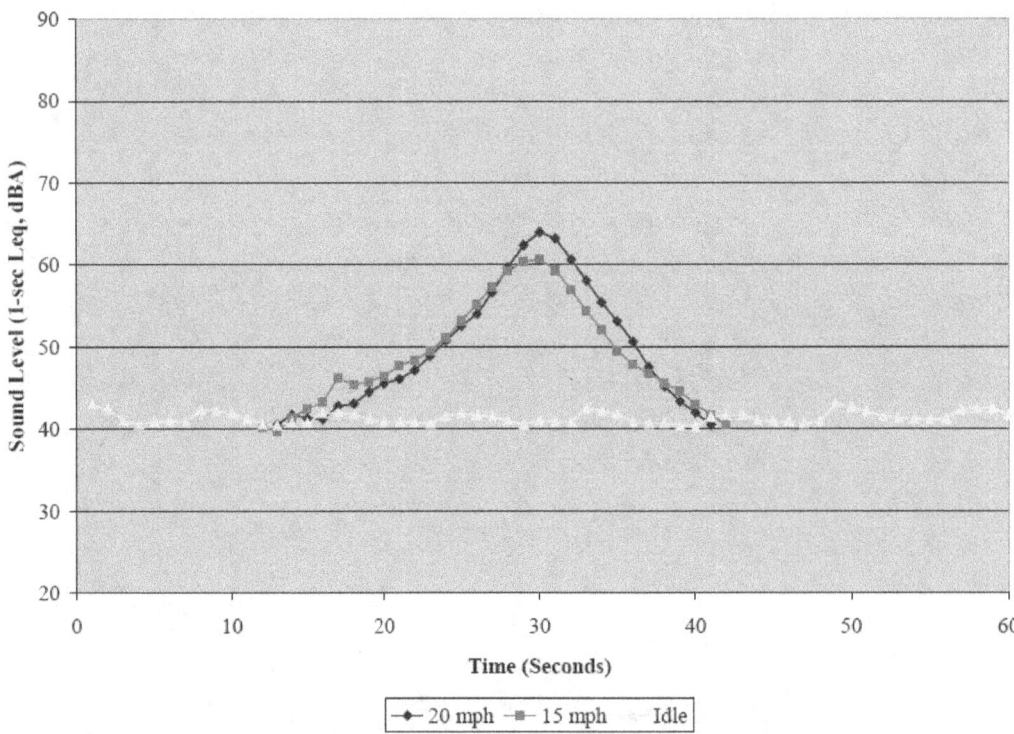

Figure 57. Yellowstone Expedition – Hayden, North Entrance, (Jan 15th) Time History Plots

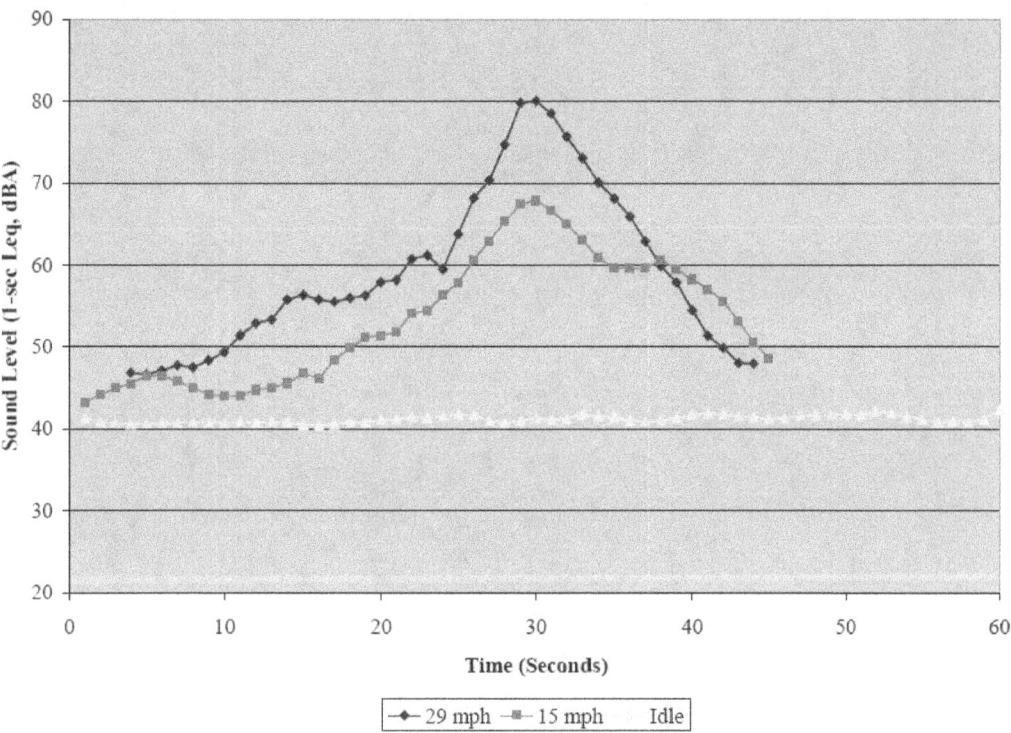

Figure 58. Xanterra 707, North Entrance, (Jan 16th) Time History Plots

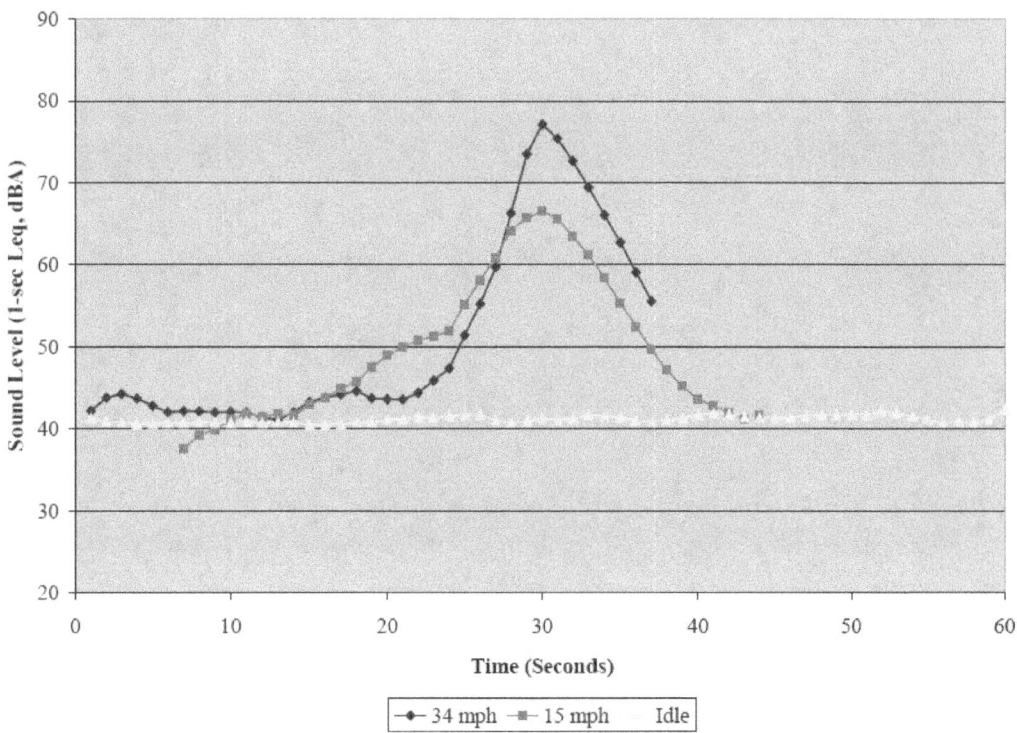

Figure 59. Yellowstone Snowcoach – SNOVAN5, North Entrance, (Jan 16th) Time History Plots

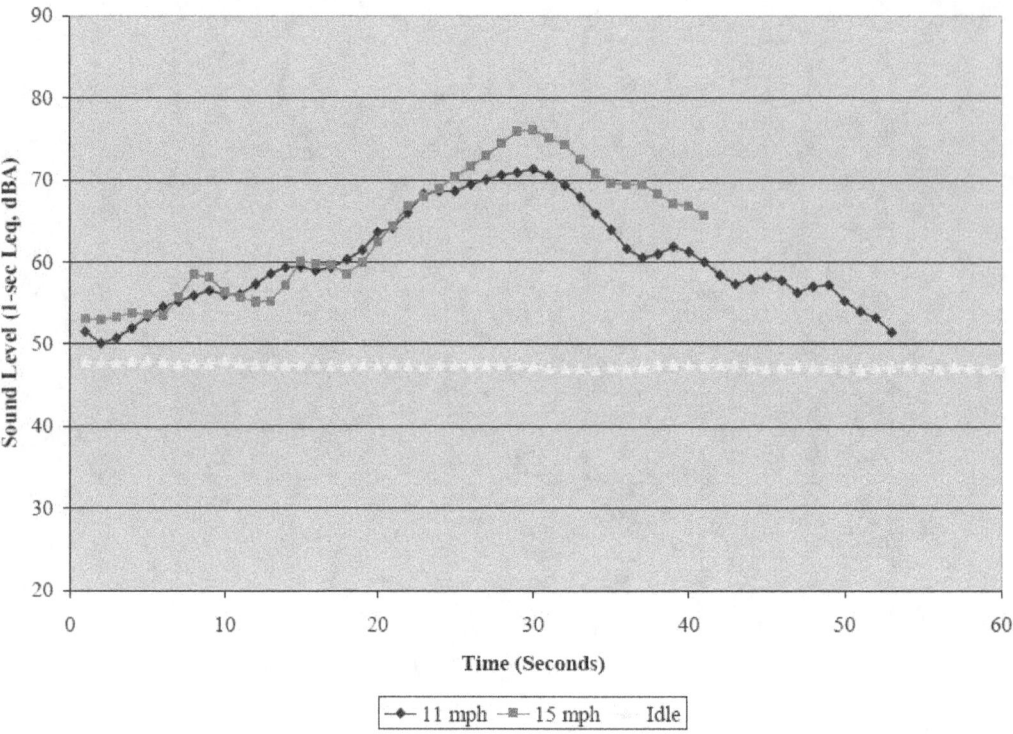

Figure 60. Xanterra 537 (Pernoth), North Entrance, (Jan 16th) Time History Plots

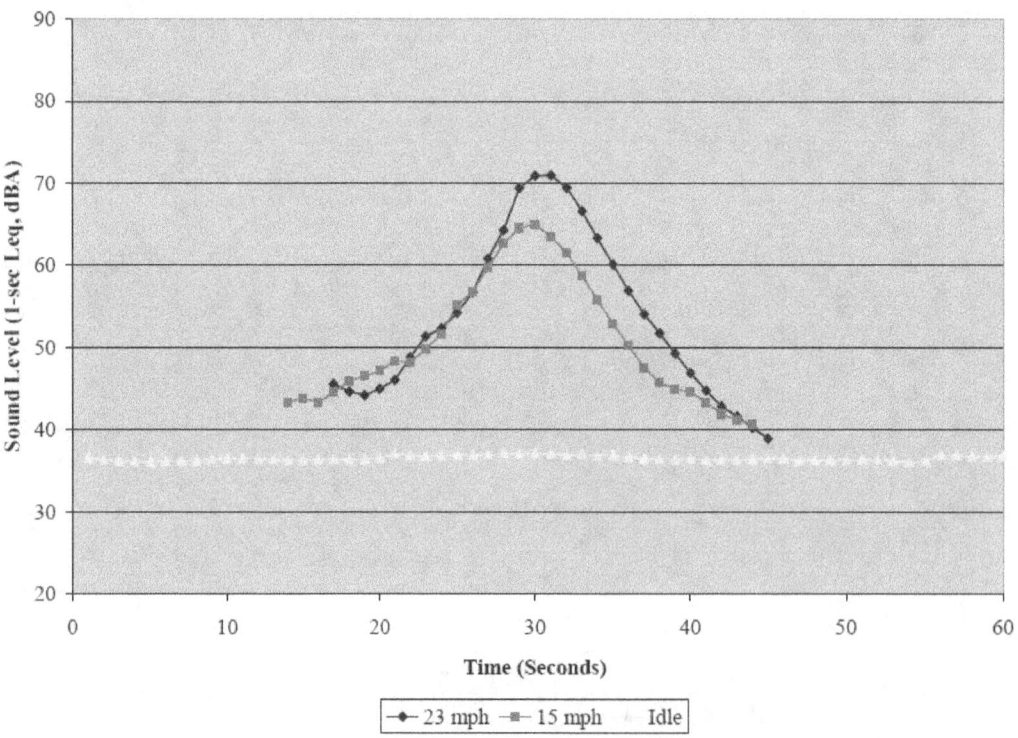

Figure 61. Xanterra 430, North Entrance, (Jan 16th) Time History Plots

West Entrance Sound Level Time Histories

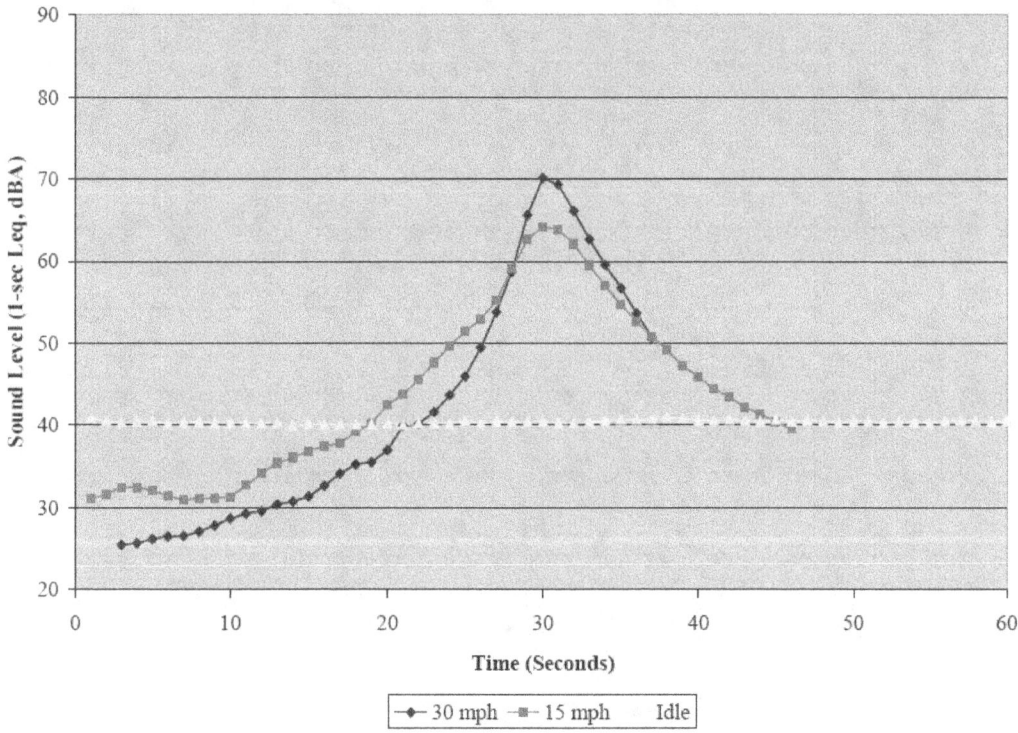

Figure 62. Yellowstone Expedition – Hayden, West Entrance, (Jan 20th) Time History Plots

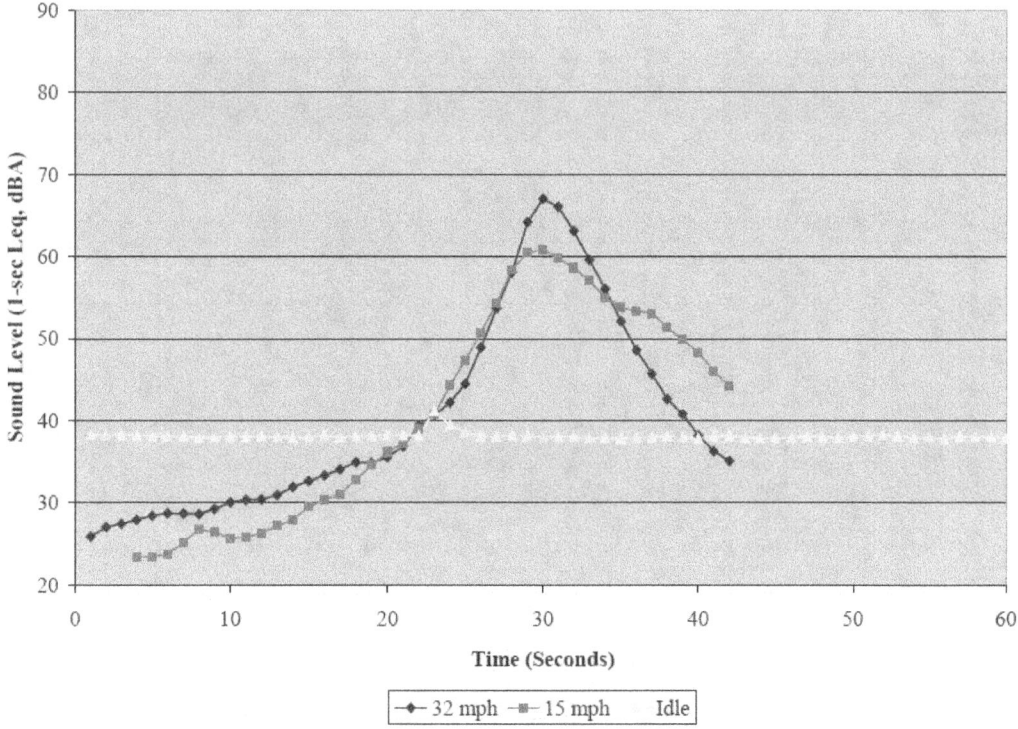

Figure 63. Alpen Guide – Kitty, West Entrance, (Jan 20th) Time History Plots

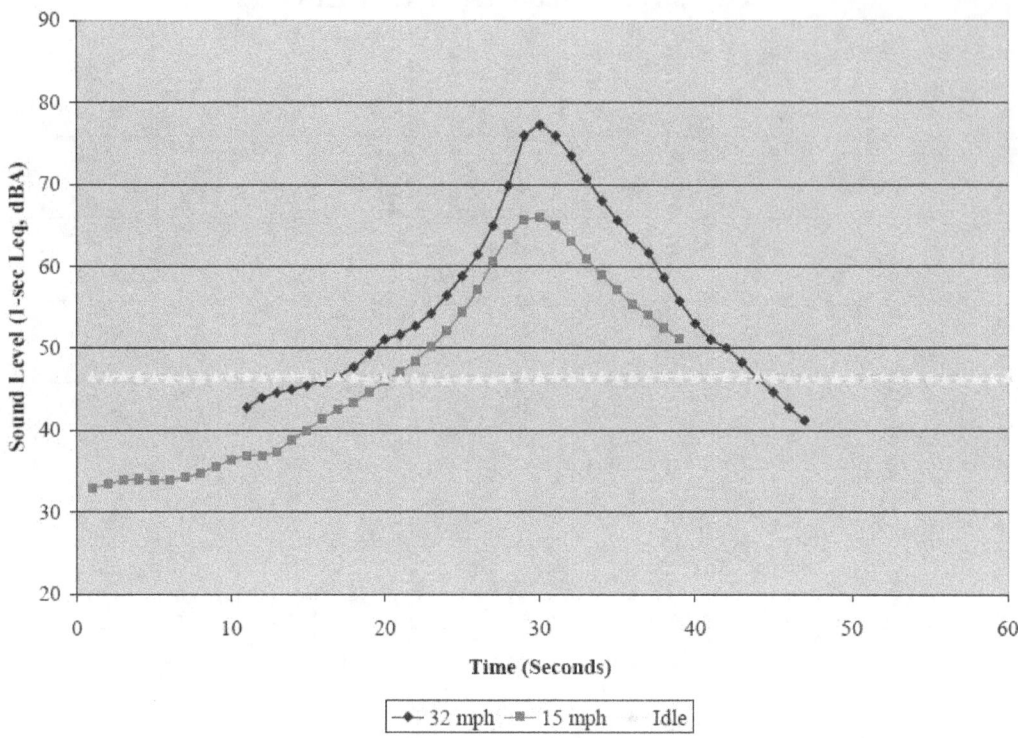

Figure 64. Xanterra 713, West Entrance, (Jan 20th) Time History Plots

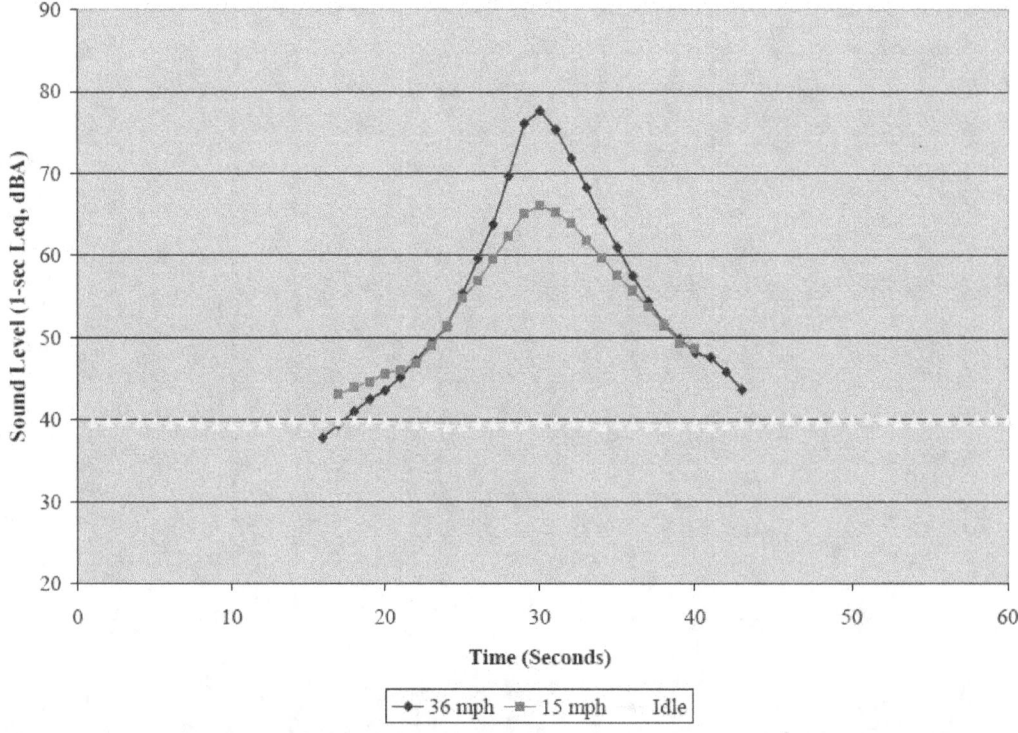

Figure 65. Yellowstone Snowcoach – SNOVAN5, West Entrance, (Jan 20th) Time History Plots

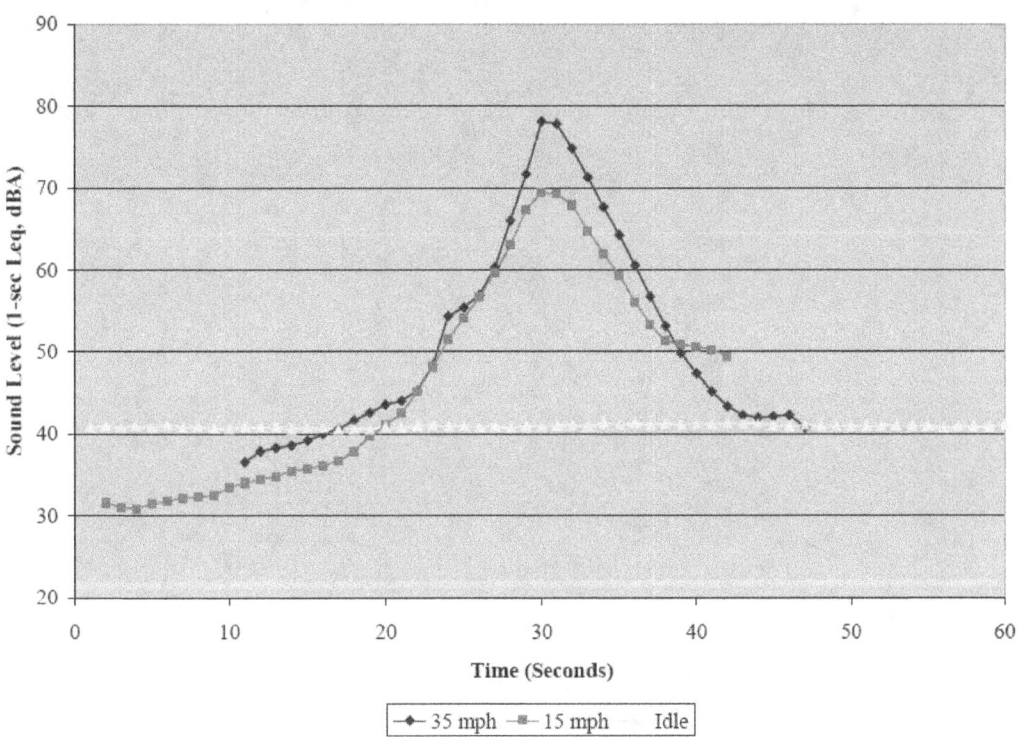

Figure 66. See Yellowstone Tours #6, West Entrance, (Jan 21st) Time History Plots

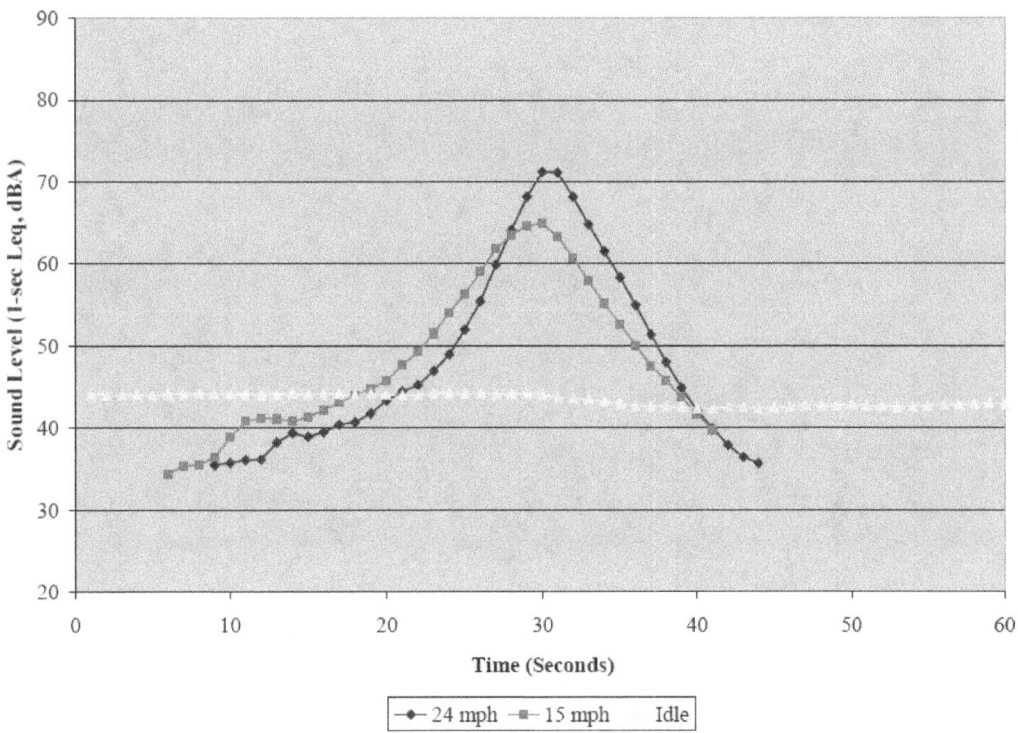

Figure 67. Buffalo Bus Touring #4, West Entrance, (Jan 21st) Time History Plots

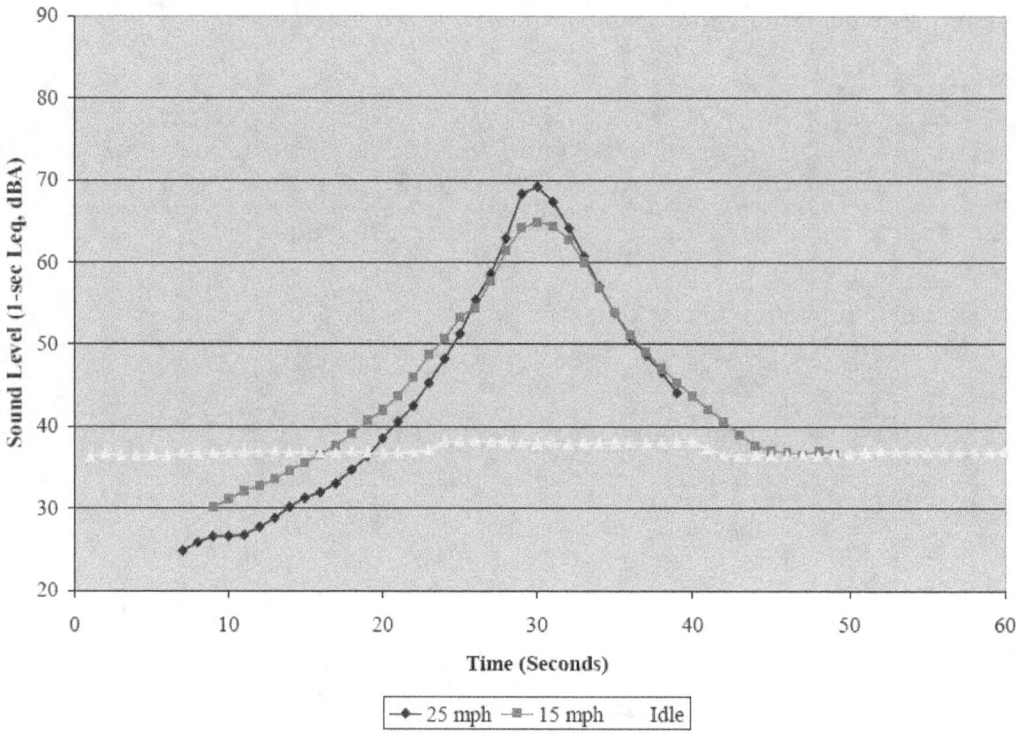

Figure 68. Yellowstone Expedition – Eleanor, West Entrance, (Jan 21st) Time History Plots

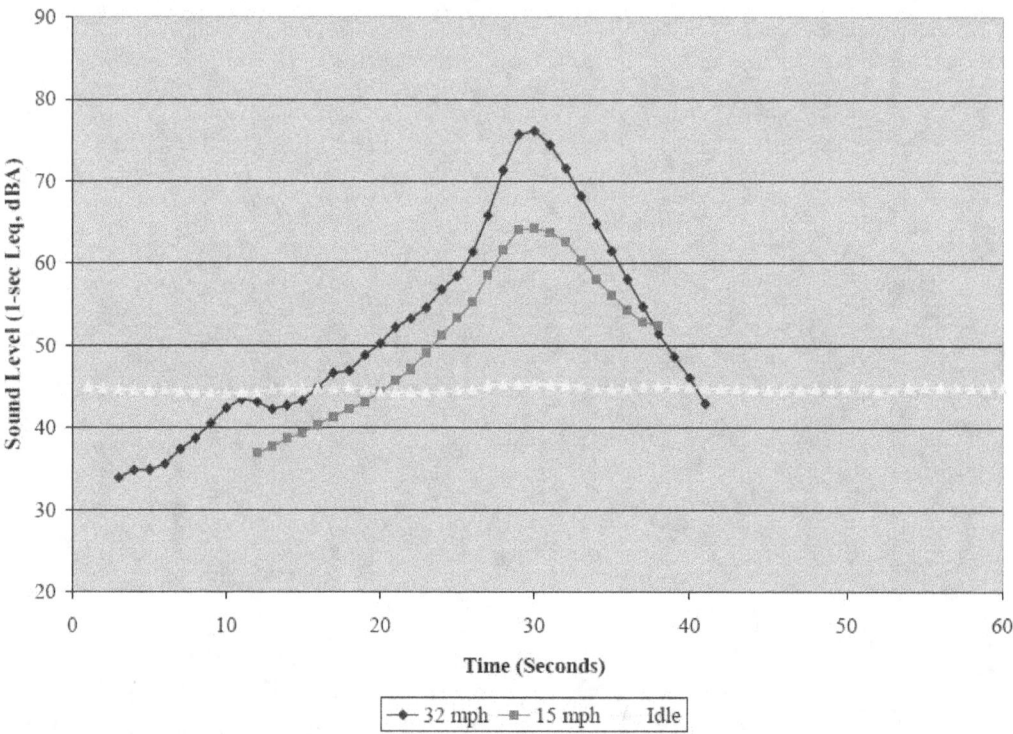

Figure 69. Xanterra 710, West Entrance, (Jan 21st) Time History Plots

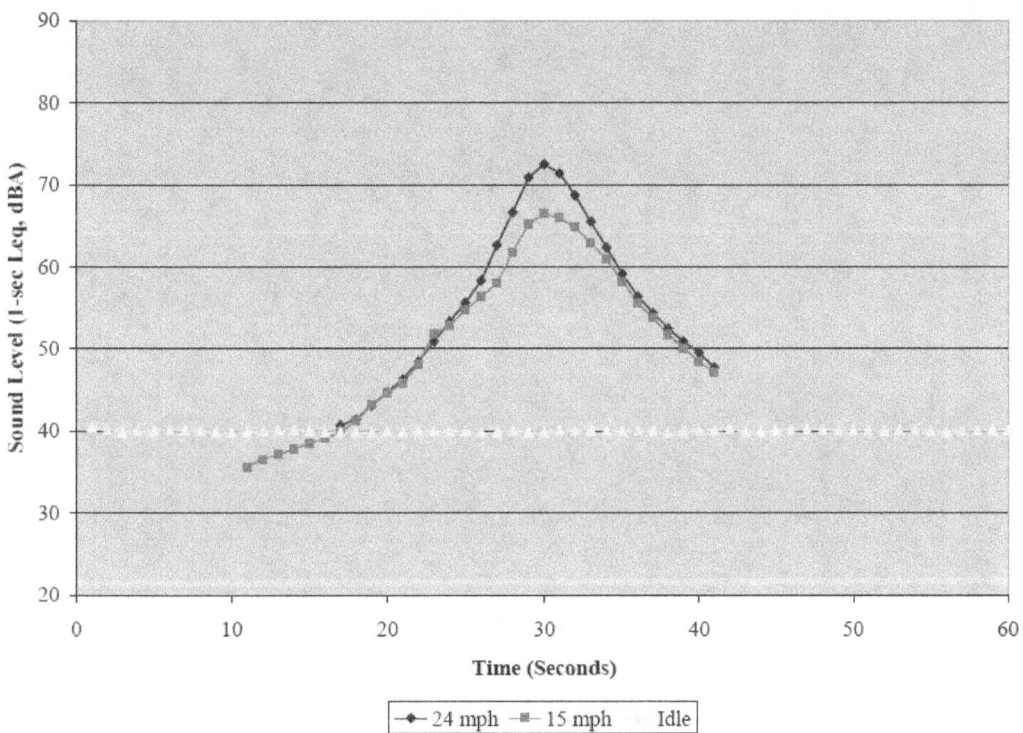

Figure 70. Buffalo Bus Touring T2, West Entrance, (Jan 21st) Time History Plots

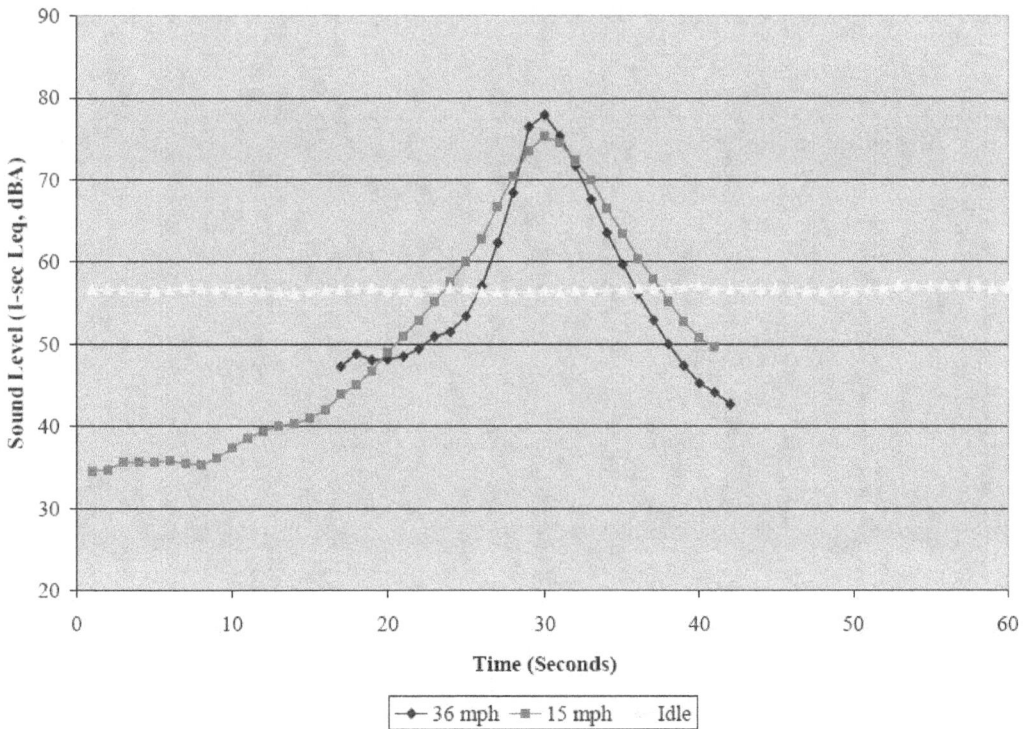

Figure 71. See Yellowstone Tours #9, West Entrance, (Jan 22nd) Time History Plots

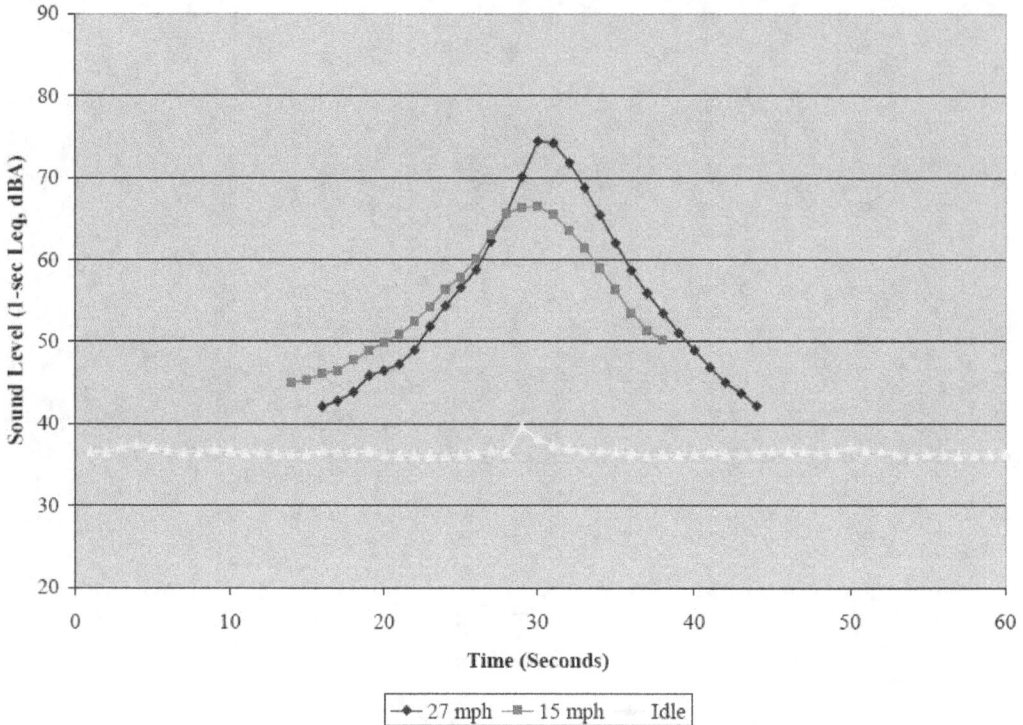

Figure 72. Buffalo Bus Touring #3, West Entrance, (Jan 22nd) Time History Plots

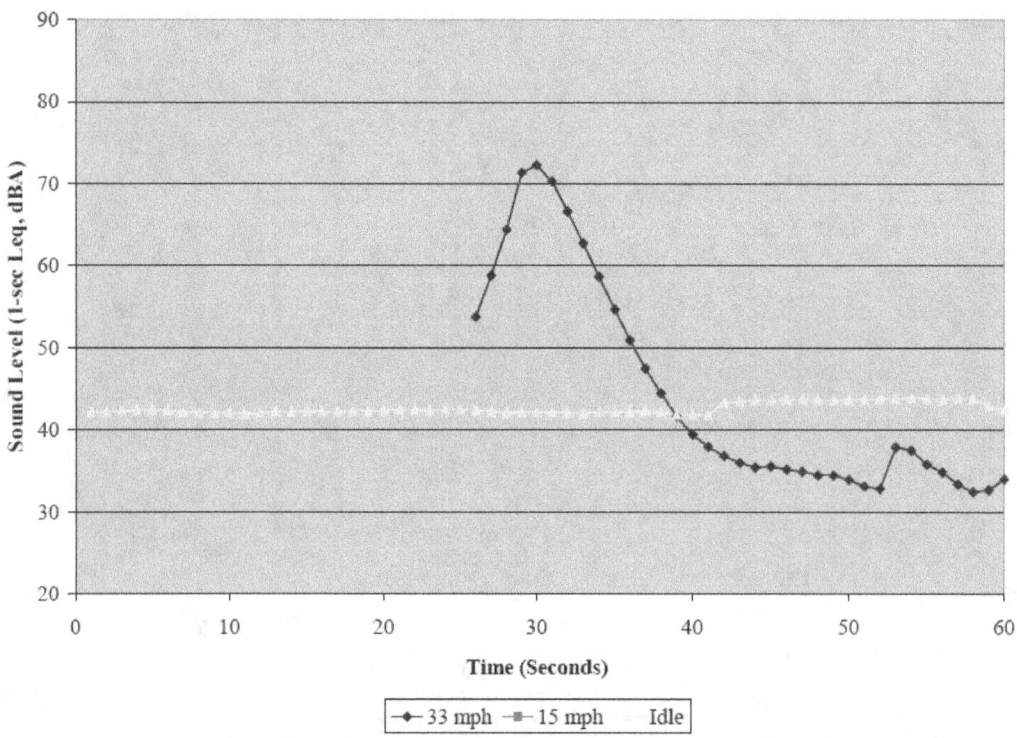

Figure 73. See Yellowstone Tours #4, West Entrance, (Jan 22nd) Time History Plots

USDOT Research & Innovative Technology Administration
Environmental Measurement and Modeling Division
Volpe Center Acoustics Facility

April 2010

Appendix E: Overall Sound Levels

Table 25 through Table 45 provide the measured, overall sound levels that were used in the determination of the L_{ASmx} / SEL for each snowcoach separated by site. Progressing from left to right data are averaged and culled, resulting in a final, single overall value. The tables are organized by speed, metric and site. Table 25 through Table 30 provide the L_{ASmx} values for the low speed events. Table 31 through Table 36 provide the L_{ASmx} values for the high speed events. Table 37 through Table 39 provide the L_{Aeq} for the idle events.

Table 40 through Table 42 provide the SEL values for the low speed events. Table 43 through Table 45 provide the SEL values for the high speed events. It should be noted that SEL calculations were only computed for the side of individual vehicles that was determined to be the loudest, based on L_{ASmx} values.

Table 25. Low Speed Measurements used to Generate Final Reported L_{ASmx} Sound Levels for the South Entrance

Vehicle	Pass-by Speed	Entrance	Vehicle Side*	Pass-by Sound Level (A-Max Slow) (dBA)	Pass-by Average (dBA)	Overall Sound Level (dBA)	Average Speed (mph)
Xanterra 713	Low Speed	South	Right	66.0	66	66	16
				67.1			
				66.0			
			Left	65.7	66		
				66.0			
				65.8			
Yellowstone Expedition – Hayden	Low Speed	South	Right	57.9	58	59	15
				58.2			
				58.8			
			Left	59.3	59		
				58.8			
				58.7			
Yellowstone Snow Coach – SNOVAN5	Low Speed	South	Right	58.4	59	59	15
				58.9			
				59.6			
			Left	58.3	59		
				58.2			
				59.1			
Alpen Guide – Kitty	Low Speed	South	Right	58.5	58	58	15
				57.7			
				58.3			
			Left	N/A	N/A		
				N/A			
				N/A			
Xanterra 709	Low Speed	South	Right	N/A	N/A	68	15
				N/A			
				N/A			
			Left	67.9	68		
				68.4			
				68.2			

* "Left/Right" indicates left/right side of vehicle from the driver's perspective.

Table 26. Low Speed Measurements used to Generate Final Reported L_{ASmx} Sound Levels for the North Entrance (part 1)

Vehicle	Pass-by Speed	Entrance	Vehicle Side*	Pass-by Sound Level (A-Max Slow) (dBA)	Pass-by Average (dBA)	Overall Sound Level (dBA)	Average Speed (mph)
Alpen Guide – Kitty	Low Speed	North	Right	N/A	N/A	59	15
				N/A			
				N/A			
			Left	58.6	59		
				59.3			
				58.8			
Yellowstone Snowcoach – SNOVAN4	Low Speed	North	Right	62.1	62	63	15
				62.4			
				62.3			
			Left	63.1	63		
				62.8			
				63.1			
Xanterra 713	Low Speed	North	Right	64.3	65	65	15
				64.6			
				64.8			
			Left	64.2	64		
				64.1			
				64.2			
Yellowstone Expedition – Hayden	Low Speed	North	Right	58.9	59	60	15
				59.3			
				59.5			
			Left	60.7	60		
				59.8			
				60.8			

* "Left/Right" indicates left/right side of vehicle from the driver's perspective.

Table 27. Low Speed Measurements used to Generate Final Reported L_{ASmx} Sound Levels for the North Entrance (part 2)

Vehicle	Pass-by Speed	Entrance	Vehicle Side*	Pass-by Sound Level (A-Max Slow) (dBA)	Pass-by Average (dBA)	Overall Sound Level (dBA)	Average Speed (mph)
Xanterra 707	Low Speed	North	Right	67.3	68	68	15
				67.9			
				67.9			
			Left	67.6	67		
				68.0			
				66.2			
Yellowstone Snowcoach – SNOVAN5	Low Speed	North	Right	64.5	65	66	16
				65.4			
				65.7			
			Left	66.5	66		
				65.1			
				65.7			
Xanterra 527	Low Speed	North	Right	70.4	71	71	11
				71.3			
				71.4			
			Left	N/A	N/A		
				N/A			
				N/A			
Xanterra 430	Low Speed	North	Right	64.8	64	65	17
				64.6			
				64.1			
			Left	65.0	65		
				65.4			
				64.0			

* "Left/Right" indicates left/right side of vehicle from the driver's perspective.

Table 28. Low Speed Measurements used to Generate Final Reported L_{ASmx} Sound Levels for the West Entrance (part 1)

Vehicle	Pass-by Speed	Entrance	Vehicle Side*	Pass-by Sound Level (A-Max Slow) (dBA)	Pass-by Average (dBA)	Overall Sound Level (dBA)	Average Speed (mph)
Yellowstone Expedition – Hayden	Low Speed	West	Right	64.3	64	64	16
				63.8			
				64.2			
			Left	63.7	64		
				63.6			
				63.8			
Alpen Guide – Kitty	Low Speed	West	Right	60.2	60	61	16
				60.7			
				60.4			
			Left	61.6	61		
				60.6			
				60.9			
Xanterra 713	Low Speed	West	Right	65.8	66	66	15
				66.4			
				66.0			
			Left	63.8	64		
				64.1			
				64.3			
Yellowstone Snowcoach – SNOVAN5	Low Speed	West	Right	65.6	66	66	15
				66.6			
				65.8			
			Left	66.6	66		
				66.0			
				65.9			

*"Left/Right" indicates left/right side of vehicle from the driver's perspective.

Table 29. Low Speed Measurements used to Generate Final Reported L_{ASmx} Sound Levels for the West Entrance (part 2)

Vehicle	Pass-by Speed	Entrance	Vehicle Side*	Pass-by Sound Level (A-Max Slow) (dBA)	Pass-by Average (dBA)	Overall Sound Level (dBA)	Average Speed (mph)
See Yellowstone Tours #6	Low Speed	West	Right	69.5	69	69	16
				69.8			
				68.9			
			Left	69.1	68		
				67.5			
				67.1			
Buffalo Bus Touring #4	Low Speed	West	Right	64.2	64	65	16
				64.0			
				63.9			
			Left	64.9	65		
				65.7			
				64.6			
Yellowstone Expedition – Eleanor	Low Speed	West	Right	63.7	64	65	15
				63.9			
				63.1			
			Left	65.0	65		
				64.9			
				64.2			
Xanterra 710	Low Speed	West	Right	64.5	64	64	15
				64.3			
				63.4			
			Left	64.7	64		
				63.8			
				64.3			

* "Left/Right" indicates left/right side of vehicle from the driver's perspective.

Table 30. Low Speed Measurements used to Generate Final Reported L_{ASmx} Sound Levels for the West Entrance (part 3)

Vehicle	Pass-by Speed	Entrance	Vehicle Side[*]	Pass-by Sound Level (A-Max Slow) (dBA)	Pass-by Average (dBA)	Overall Sound Level (dBA)	Average Speed (mph)
Buffalo Bus Touring #T2	Low Speed	West	Right	66.1	66	67	16
				65.6			
				65.0			
			Left	66.7	67		
				66.4			
				66.5			
See Yellowstone #9	Low Speed	West	Right	75.3	76	76	15
				76.4			
				75.2			
			Left	73.1	72		
				72.1			
				71.6			
Buffalo Bus Touring #3	Low Speed	West	Right	66.4	67	67	15
				66.5			
				66.8			
			Left	N/A	N/A		
				N/A			
				N/A			
See Yellowstone Tours #4	Low Speed	West	Right	N/A[**]	N/A[**]	N/A[**]	N/A[**]
				N/A[**]			
				N/A[**]			
			Left	N/A[**]	N/A[**]		
				N/A[**]			
				N/A[**]			

* "Left/Right" indicates left/right side of vehicle from the driver's perspective.
** See Yellowstone Tours #4 did not have a low speed series due to inclimate weather.

Table 31. High Speed Measurements used to Generate Final Reported L_{ASmx} Sound Levels for the South Entrance

Vehicle	Pass-by Speed	Entrance	Vehicle Side*	Pass-by Sound Level (A-Max Slow) (dBA)	Pass-by Average (dBA)	Overall Sound Level (dBA)	Average Speed (mph)
Xanterra 713	High Speed	South	Right	74.8	75	75	23
				74.8			
				75.5			
			Left	74.6	75		
				76.4			
				74.6			
Yellowstone Expedition – Hayden	High Speed	South	Right	62.2	63	63	20
				62.3			
				63.1			
			Left	63.4	63		
				63.4			
				63.5			
Yellowstone Snow Coach – SNOVAN5	High Speed	South	Right	67.7	68	68	25
				67.4			
				68.7			
			Left	67.2	68		
				68.2			
				68.1			
Alpen Guide – Kitty	High Speed	South	Right	66.0	66	66	32
				66.5			
				66.2			
			Left	N/A	N/A		
				N/A			
				N/A			
Xanterra 709	High Speed	South	Right	75.9	76	80	28
				75.8			
				76.3			
			Left	79.2	80		
				80.1			
				80.0			

* "Left/Right" indicates left/right side of vehicle from the driver's perspective.

Table 32. High Speed Measurements used to Generate Final Reported L_{ASmx} Sound Levels for the North Entrance (part 1)

Vehicle	Pass-by Speed	Entrance	Vehicle Side*	Pass-by Sound Level (A-Max Slow) (dBA)	Pass-by Average (dBA)	Overall Sound Level (dBA)	Average Speed (mph)
Alpen Guide – Kitty	High Speed	North	Right	67.0	67	67	29
				66.9			
				66.3			
			Left	67.6	66		
				65.8			
				65.9			
Yellowstone Snowcoach – SNOVAN4	High Speed	North	Right	76.2	76	76	33
				75.2			
				75.5			
			Left	75.1	75		
				74.2			
				74.9			
Xanterra 713	High Speed	North	Right	75.6	76	76	26
				75.8			
				75.2			
			Left	75.3	75		
				75.2			
				74.8			
Yellowstone Expedition – Hayden	High Speed	North	Right	63.3	63	64	20
				62.6			
				62.0			
			Left	64.8	64		
				63.6			
				64.0			

* "Left/Right" indicates left/right side of vehicle from the driver's perspective.

Table 33. High Speed Measurements used to Generate Final Reported L_{ASmx} Sound Levels for the North Entrance (part 2)

Vehicle	Pass-by Speed	Entrance	Vehicle Side*	Pass-by Sound Level (A-Max Slow) (dBA)	Pass-by Average (dBA)	Overall Sound Level (dBA)	Average Speed (mph)
Xanterra 707	High Speed	North	Right	77.1	76	80	29
				75.6			
				75.6			
			Left	80.5	80		
				79.6			
				80.0			
Yellowstone Snowcoach – SNOVAN5	High Speed	North	Right	77.0	77	77	34
				77.2			
				77.2			
			Left	76.6	76		
				76.5			
				75.7			
Xanterra 537	High Speed	North	Right	75.6	76	76	16
				76.3			
				74.9			
			Left	77.3	76		
				76.0			
				76.1			
Xanterra 430	High Speed	North	Right	70.3	69	70	23
				68.9			
				69.2			
			Left	71.0	70		
				71.2			
				69.3			

* "Left/Right" indicates left/right side of vehicle from the driver's perspective.

Table 34. High Speed Measurements used to Generate Final Reported L_{ASmx} Sound Levels for the West Entrance (part 1)

Vehicle	Pass-by Speed	Entrance	Vehicle Side*	Pass-by Sound Level (A-Max Slow) (dBA)	Pass-by Average (dBA)	Overall Sound Level (dBA)	Average Speed (mph)
Yellowstone Expedition – Hayden	High Speed	West	Right	69.7	70	70	30
				69.4			
				70.3			
			Left	68.9	70		
				70.7			
				70.2			
Alpen Guide – Kitty	High Speed	West	Right	67.1	67	67	32
				67.2			
				66.6			
			Left	66.3	67		
				66.7			
				67.1			
Xanterra 713	High Speed	West	Right	77.3	77	77	32
				77.1			
				77.8			
			Left	77.0	77		
				76.9			
				76.1			
Yellowstone Snowcoach – SNOVAN5	High Speed	West	Right	76.4	77	78	36
				77.6			
				76.5			
			Left	78.6	78		
				77.7			
				77.6			

*"Left/Right" indicates left/right side of vehicle from the driver's perspective.

Table 35. High Speed Measurements used to Generate Final Reported L$_{ASmx}$ Sound Levels for the West Entrance (part 2)

Vehicle	Pass-by Speed	Entrance	Vehicle Side*	Pass-by Sound Level (A-Max Slow) (dBA)	Pass-by Average (dBA)	Overall Sound Level (dBA)	Average Speed (mph)
See Yellowstone Tours #6	High Speed	West	Right	78.1	78	78	35
				77.1			
				78.2			
			Left	77.1	77		
				77.2			
				77.1			
Buffalo Bus Touring #4	High Speed	West	Right	67.4	68	71	24
				68.4			
				68.3			
			Left	71.2	71		
				72.0			
				71.2			
Yellowstone Expedition – Eleanor	High Speed	West	Right	68.1	68	68	25
				68.1			
				67.3			
			Left	69.2	68		
				68.0			
				68.2			
Xanterra 710	High Speed	West	Right	76.3	76	76	32
				76.2			
				76.0			
			Left	76.3	76		
				75.7			
				75.3			

* "Left/Right" indicates left/right side of vehicle from the driver's perspective.

Table 36. High Speed Measurements used to Generate Final Reported L_{ASmx} Sound Levels for the West Entrance (part 3)

Vehicle	Pass-by Speed	Entrance	Vehicle Side*	Pass-by Sound Level (A-Max Slow) (dBA)	Pass-by Average (dBA)	Overall Sound Level (dBA)	Average Speed (mph)
Buffalo Bus Touring #T2	High Speed	West	Right	71.8	72	72	24
				71.5			
				71.6			
			Left	72.6	72		
				72.5			
				71.9			
See Yellowstone #9	High Speed	West	Right	76.2	77	78	36
				77.8			
				77.9			
			Left	77.5	78		
				77.9			
				78.1			
Buffalo Bus Touring #3	High Speed	West	Right	73.7	74	75	27
				73.9			
				73.6			
			Left	74.5	75		
				74.4			
				74.8			
See Yellowstone Tours #4	High Speed	West	Right	72.3	72	72	33
				72.2			
				72.6			
			Left	71.1	71		
				70.1			
				71.6			

* "Left/Right" indicates left/right side of vehicle from the driver's perspective.

Table 37. Idle Measurements to Generate Final Reported L_{Aeq} Sound Levels for the South Entrance

Vehicle	Pass-by Speed	Entrance	Vehicle Side*	Pass-by Sound Level (Average) (dBA)	Overall Sound Level (dBA)
Xanterra 713	Idle	South	Right	42.2	43
			Left	43.2	
Yellowstone Expedition – Hayden	Idle	South	Right	37.4	39
			Left	38.5	
Yellowstone Snowcoach – SNOVAN5	Idle	South	Right	38.3	38
			Left	37.3	
Alpen Guide – Kitty	Idle	South	Right	38.9	39
			Left	36.5	
Xanterra 709	Idle	South	Right	48.6	49
			Left	49	

* "Left/Right" indicates left/right side of vehicle from the driver's perspective.

Table 38. Idle Measurements to Generate Final Reported L_{Aeq} Sound Levels for the North Entrance

Vehicle	Pass-by Speed	Entrance	Vehicle Side*	Pass-by Sound Level (Average) (dBA)	Overall Sound Level (dBA)
Alpen Guide – Kitty	Idle	North	Right	40.9	41
			Left	40.1	
Yellowstone Snowcoach – SNOVAN4	Idle	North	Right	40.4	41
			Left	40.9	
Xanterra 713	Idle	North	Right	44.2	45
			Left	45	
Yellowstone Expedition – Hayden	Idle	North	Right	38.2	41
			Left	41.4	
Xanterra 707	Idle	North	Right	47	47
			Left	47.1	
Yellowstone Snowcoach – SNOVAN5	Idle	North	Right	41.2	41
			Left	40.6	
Xanterra 537	Idle	North	Right	47.1	47
			Left	47.3	
Xanterra 430	Idle	North	Right	36.5	37
			Left	35.2	

* "Left/Right" indicates left/right side of vehicle from the driver's perspective.

Table 39. Idle Measurements to Generate Final Reported L$_{Aeq}$ Sound Levels for the West Entrance

Vehicle	Pass-by Speed	Entrance	Vehicle Side*	Pass-by Sound Level (Average) (dBA)	Overall Sound Level (dBA)
Yellowstone Expedition – Hayden	Idle	West	Right	37.7	40
			Left	40.4	
Alpen Guide – Kitty	Idle	West	Right	38.2	38
			Left	37.4	
Xanterra 713	Idle	West	Right	45.6	46
			Left	46.4	
Yellowstone Snowcoach – SNOVAN5	Idle	West	Right	39.4	40
			Left	39.6	
See Yellowstone Tours #6	Idle	West	Right	40.9	41
			Left	39.6	
Buffalo Bus Touring #4	Idle	West	Right	43.5	44
			Left	43.2	
Yellowstone Expedition – Eleanor	Idle	West	Right	37.1	37
			Left	37	
Xanterra 710	Idle	West	Right	44.7	45
			Left	44.1	
Buffalo Bus Touring T2	Idle	West	Right	40	40
			Left	39.7	
See Yellowstone Tours #9	Idle	West	Right	56.3	57
			Left	56.7	
Buffalo Bus Touring #3	Idle	West	Right	36.6	37
			Left	36.6	
Buffalo Bus Touring #4	Idle	West	Right	42.7	43
			Left	41.7	

* "Left/Right" indicates left/right side of vehicle from the driver's perspective.

Table 40. Low Speed Measurements used to Generate Final Reported SEL Sound Levels for the South Entrance

Vehicle	Pass-by Speed	Entrance	Vehicle Side*	Pass-by SEL Sound Level (dBA)	Overall Sound Level (dBA)	Average Speed (mph)
Xanterra 713	Low Speed	South	Right	72.7	73	16
				73.5		
				72.9		
Yellowstone Expedition – Hayden	Low Speed	South	Left	66.1	66	15
				65.4		
				65.6		
Yellowstone Snowcoach – SNOVAN5	Low Speed	South	Right	65.6	66	15
				65.9		
				66.6		
Alpen Guide – Kitty	Low Speed	South	Right	64.4	64	15
				63.9		
				64.2		
Xanterra 709	Low Speed	South	Left	75.1	75	15
				75.6		
				75.1		

*"Left/Right" indicates left/right side of vehicle from the driver's perspective.

Table 41. Low Speed Measurements used to Generate Final Reported SEL Sound Levels for the North Entrance

Vehicle	Pass-by Speed	Entrance	Vehicle Side[*]	Pass-by SEL Sound Level (dBA)	Overall Sound Level (dBA)	Average Speed (mph)
Alpen Guide – Kitty	Low Speed	North	Left	66.1	66	15
				66.2		
				66.2		
Yellowstone Snowcoach – SNOVAN4	Low Speed	North	Left	70.4	70	15
				69.8		
				70.0		
Xanterra 713	Low Speed	North	Right	72.7	73	15
				72.0		
				73.2		
Yellowstone Expedition – Hayden	Low Speed	North	Left	68.0	68	15
				66.9		
				67.8		
Xanterra 707	Low Speed	North	Left	75.6	76	15
				77.4		
				75.0		
Yellowstone Snowcoach – SNOVAN5	Low Speed	North	Left	73.3	73	16
				72.1		
				72.4		
Xanterra 537	Low Speed	North	Right	80.9	81	11
				80.9		
				81.8		
Xanterra 430	Low Speed	North	Right	72.5	77	23
				72.0		
				71.5		

[*] "Left/Right" indicates left/right side of vehicle from the driver's perspective.

Table 42. Low Speed Measurements used to Generate Final Reported SEL Sound Levels for the West Entrance

Vehicle	Pass-by Speed	Entrance	Vehicle Side*	Pass-by SEL Sound Level (dBA)	Overall Sound Level (dBA)	Average Speed (mph)
Yellowstone Expedition – Hayden	Low Speed	West	Right	71.0 70.5 70.7	71	16
Alpen Guide – Kitty	Low Speed	West	Left	68.5 68.1 68.4	68	16
Xanterra 713	Low Speed	West	Right	73.1 73.6 73.1	73	15
Yellowstone Snowcoach – SNOVAN5	Low Speed	West	Right	72.9 73.8 73.3	73	15
See Yellowstone Tours #6	Low Speed	West	Right	75.8 76.0 75.8	76	16
Buffalo Bus Touring #4	Low Speed	West	Left	72.0 72.5 71.8	72	16
Yellowstone Expedition – Eleanor	Low Speed	West	Left	71.3 71.6 71.0	71	15
Xanterra 710	Low Speed	West	Left	72.4 71.3 71.7	72	15
Buffalo Bus Touring T2	Low Speed	West	Left	73.8 73.5 73.3	74	16
See Yellowstone Tours #9	Low Speed	West	Right	81.4 81.6 81.2	81	15
Buffalo Bus Touring #3	Low Speed	West	Left	74.1 74.0 74.2	74	15
See Yellowstone Tours #4	Low Speed	West	N/A	N/A N/A N/A	N/A	N/A

* "Left/Right" indicates left/right side of vehicle from the driver's perspective.

Table 43. High Speed Measurements used to Generate Final Reported SEL Sound Levels for the South Entrance

Vehicle	Pass-by Speed	Entrance	Vehicle Side*	Pass-by SEL Sound Level (dBA)	Overall Sound Level (dBA)	Average Speed (mph)
Xanterra 713	High Speed	South	Left	81.1	81	23
				82.5		
				80.8		
Yellowstone Expedition – Hayden	High Speed	South	Left	69.2	69	20
				69.1		
				69.3		
Yellowstone Snowcoach – SNOVAN5	High Speed	South	Right	73.6	73	25
				72.8		
				73.7		
Alpen Guide – Kitty	High Speed	South	Right	70.2	71	32
				71.2		
				70.4		
Xanterra 709	High Speed	South	Left	84.1	84	28
				84.2		
				85.0		

* "Left/Right" indicates left/right side of vehicle from the driver's perspective.

Table 44. High Speed Measurements used to Generate Final Reported SEL Sound Levels for the North Entrance

Vehicle	Pass-by Speed	Entrance	Vehicle Side*	Pass-by SEL Sound Level (dBA)	Overall Sound Level (dBA)	Average Speed (mph)
Alpen Guide – Kitty	High Speed	North	Right	72.6	72	29
				72.3		
				71.1		
Yellowstone Snowcoach – SNOVAN4	High Speed	North	Right	80.7	80	33
				79.9		
				80.0		
Xanterra 713	High Speed	North	Right	81.5	82	26
				81.9		
				81.2		
Yellowstone Expedition – Hayden	High Speed	North	Left	71.1	71	20
				70.4		
				70.1		
Xanterra 707	High Speed	North	Left	85.7	85	29
				84.8		
				85.9		
Yellowstone Snowcoach – SNOVAN5	High Speed	North	Right	81.6	82	34
				81.8		
				81.6		
Xanterra 537	High Speed	North	Left	85.5	85	16
				84.4		
				84.9		
Xanterra 430	High Speed	North	Left	77.3	77	23
				77.1		
				75.3		

*"Left/Right" indicates left/right side of vehicle from the driver's perspective.

Table 45. High Speed Measurements used to Generate Final Reported SEL Sound Levels for the West Entrance

Vehicle	Pass-by Speed	Entrance	Vehicle Side*	Pass-by SEL Sound Level (dBA)	Overall Sound Level (dBA)	Average Speed (mph)
Yellowstone Expedition – Hayden	High Speed	West	Left	74.2	75	30
				75.2		
				74.8		
Alpen Guide – Kitty	High Speed	West	Right	72.1	72	32
				72.3		
				71.7		
Xanterra 713	High Speed	West	Right	82.8	83	32
				82.7		
				83.3		
Yellowstone Snowcoach – SNOVAN5	High Speed	West	Left	82.7	82	36
				82.3		
				82.2		
See Yellowstone Tours #6	High Speed	West	Right	82.9	82	35
				81.2		
				82.9		
Buffalo Bus Touring #4	High Speed	West	Left	76.4	77	24
				77.1		
				76.8		
Yellowstone Expedition – Eleanor	High Speed	West	Left	74.4	74	25
				73.5		
				73.2		
Xanterra 710	High Speed	West	Left	81.6	82	32
				81.7		
				81.5		
Buffalo Bus Touring T2	High Speed	West	Left	78.1	78	24
				78.1		
				77.3		
See Yellowstone Tours #9	High Speed	West	Left	82.1	82	36
				82.3		
				82.0		
Buffalo Bus Touring #3	High Speed	West	Left	79.9	80	27
				79.6		
				79.9		
See Yellowstone Tours #4	High Speed	West	Right	77.2	77	33
				76.9		
				76.7		

* "Left/Right" indicates left/right side of vehicle from the driver's perspective.

This page intentionally left blank

USDOT Research & Innovative Technology Administration
Environmental Measurement and Modeling Division
Volpe Center Acoustics Facility

April 2010

Appendix F: One-Third Octave Band Sound Levels

One-third octave band sound levels associated with the high and low speed pass-by time histories found in Appendix E are provided by site. One-third octave band sound levels are flat and occur at the time of L_{ASmx}. One-third octave band sound levels associated with the south entrance are represented in Table 46 and Table 47, and graphically in Figure 74 to Figure 78. One-third octave band sound levels associated with the north entrance are represented in Table 48 and Table 49, and graphically in Figure 79 to Figure 86. One-third octave band sound levels associated with the west entrance are represented in Table 50 though Table 52, and graphically in Figure 87 though Figure 98.

Table 46. Maximum One-Third Octave Band Sound Levels for Select South Entrance Events (part 1), dB

Vehicle		Xanterra 713	Xanterra 713	Yellowstone Expedition – Hayden	Yellowstone Expedition – Hayden	Yellowstone Snowcoach – SNOVAN5	Yellowstone Snowcoach – SNOVAN5
Speed		High	Low	High	Low	High	Low
Vehicle Side*		Left	Right	Left	Left	Right	Right
Max Time		15:30:07	16:03:04	14:21:19	14:49:30	13:28:57	13:55:49
		Maximum Sound Pressure Level (dBA)					
One-Third Octave Band Center Frequency, (Hz)	12.5	56.3	46.7	52.6	53.6	61.2	59.3
	16	57.3	59.9	57.3	51.2	64.6	61.3
	20	62.8	60.1	54.0	56.0	64.1	58.6
	25	69.6	63.5	59.5	56.0	57.8	55.6
	31.5	60.6	69.8	58.7	55.2	57.6	52.0
	40	59.9	60.2	58.9	57.3	57.7	54.5
	50	72.0	64.7	60.2	62.9	59.7	56.1
	63	69.4	61.5	65.1	69.1	56.6	60.6
	80	67.0	71.3	66.8	63.5	55.5	71.9
	100	74.8	61.0	59.7	57.7	59.6	56.5
	125	69.6	60.1	59.9	65.6	68.7	53.0
	160	64.9	61.4	63.2	58.3	61.5	60.2
	200	74.0	65.2	65.1	58.2	56.9	56.2
	250	59.7	63.7	56.0	53.7	65.9	57.1
	315	60.9	57.0	53.1	49.6	60.7	54.9
	400	66.0	58.4	52.3	51.5	61.3	51.3
	500	67.9	56.3	52.2	48.9	55.8	48.8
	630	67.2	56.9	50.7	45.9	54.8	49.8
	800	63.2	55.9	51.6	45.8	54.3	47.6
	1000	63.1	54.6	52.9	46.8	54.3	47.4
	1250	65.9	55.4	51.0	45.4	53.6	45.7
	1600	65.0	54.6	51.6	46.2	52.7	45.0
	2000	61.1	52.1	52.3	46.1	52.8	44.5
	2500	60.2	51.4	51.9	46.0	50.5	42.4
	3150	58.6	49.8	50.4	45.0	48.8	41.0
	4000	56.0	48.3	48.3	42.6	47.4	40.1
	5000	55.5	50.1	47.8	41.6	47.3	40.6
	6300	54.2	49.6	47.4	40.2	46.7	39.9
	8000	53.1	46.9	44.7	37.9	45.4	38.6
	10000	51.3	46.3	42.2	35.6	44.3	37.4
	12500	48.0	51.0	40.2	33.9	43.0	35.4
	16000	45.5	49.1	38.3	33.3	42.8	34.5
	20000	42.7	42.1	37.6	31.5	41.9	33.5

* "Left/Right" indicates left/right side of vehicle from the driver's perspective.

Table 47. Maximum One-Third Octave Band Sound Levels for Select South Entrance Events (part 2), dB

Vehicle	Alpen Guide – Kitty	Alpen Guide – Kitty	Xanterra 709	Xanterra 709
Speed	High	Low	High	Low
Vehicle Side*	Right	Right	Left	Left
Max Time	12:43:00	13:00:02	11:45:16	12:10:37
One-Third Octave Band Center Frequency (Hz)	Maximum Sound Pressure Level (dBA)			
12.5	51.0	48.9	55.9	44.9
16	55.6	54.5	60.1	60.7
20	63.9	62.9	62.4	64.0
25	65.5	63.8	77.0	70.8
31.5	61.7	61.8	75.8	67.6
40	61.3	62.0	69.6	64.8
50	62.8	58.9	70.8	66.3
63	62.7	66.0	71.0	64.0
80	60.7	70.8	73.1	74.2
100	61.7	66.3	65.8	63.6
125	63.2	65.2	72.9	62.5
160	68.2	55.0	70.4	62.0
200	56.6	52.4	68.2	61.3
250	55.2	50.6	71.8	59.3
315	55.2	47.5	71.7	56.9
400	51.3	44.9	71.4	57.8
500	53.4	46.3	70.6	55.8
630	55.4	47.3	68.3	56.9
800	55.1	45.9	67.6	57.3
1000	54.2	48.5	66.0	56.8
1250	59.4	48.1	69.5	58.0
1600	56.2	45.7	68.7	55.8
2000	55.7	43.6	67.4	55.5
2500	55.3	43.2	67.7	53.7
3150	52.6	42.9	67.5	53.6
4000	51.0	39.8	68.2	55.8
5000	48.7	38.3	69.3	58.8
6300	47.3	39.7	68.6	56.7
8000	44.9	34.7	65.5	55.4
10000	42.4	33.4	58.9	49.3
12500	39.0	31.3	54.0	43.8
16000	37.4	30.7	50.1	40.5
20000	35.2	30.1	44.0	35.7

* "Left/Right" indicates left/right side of vehicle from the driver's perspective.

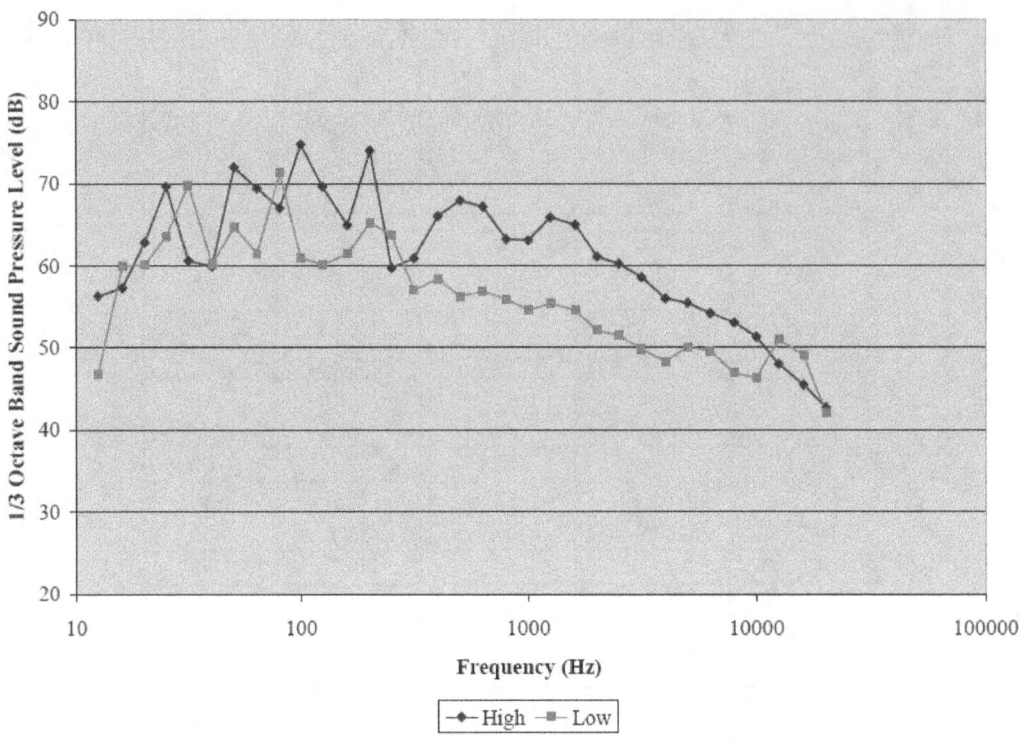

Figure 74. Xanterra 713, South Entrance (Jan 14th) Maximum Spectra for Low Speed and High Speed

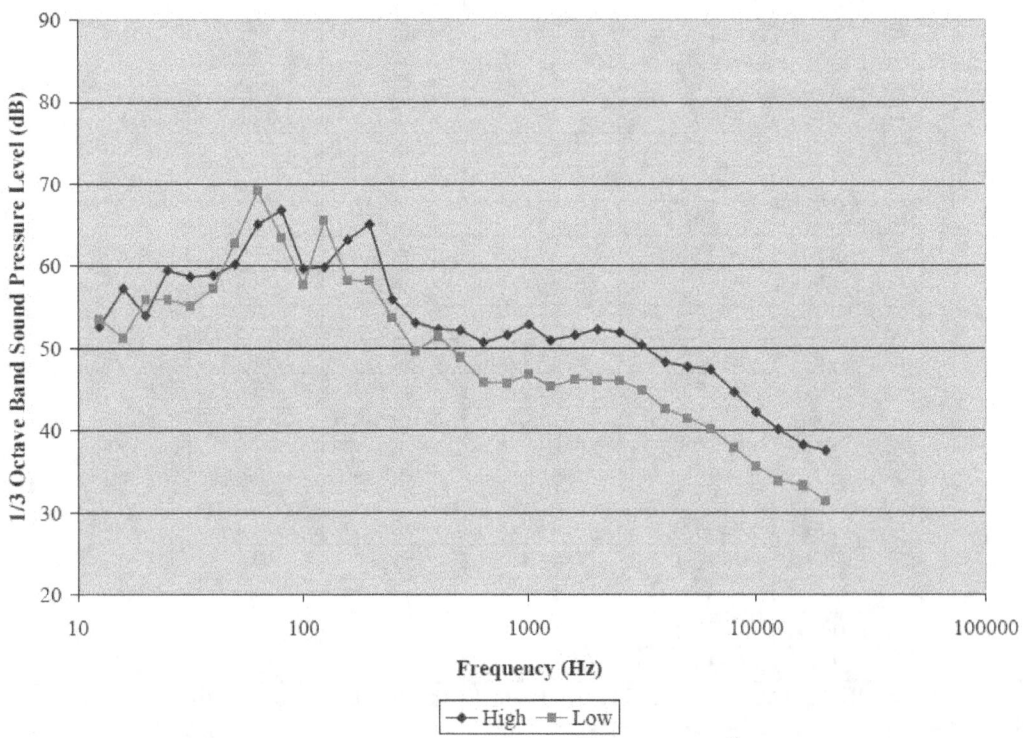

Figure 75. Yellowstone Expedition – Hayden, South Entrance (Jan 14th) Maximum Spectra for Low Speed and High Speed

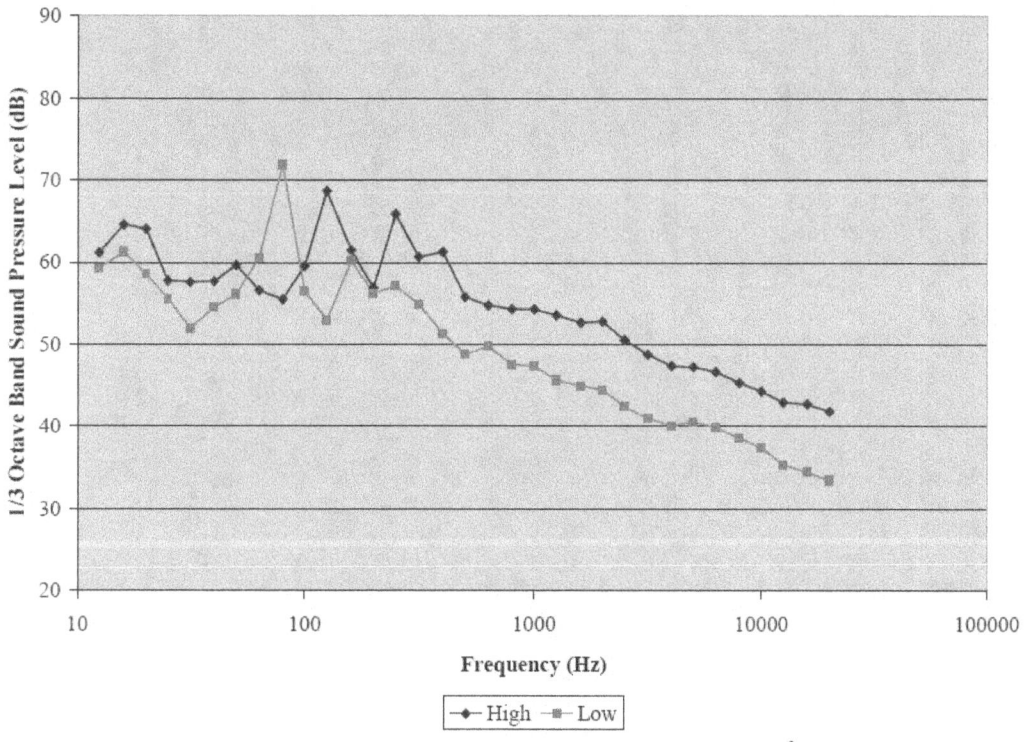

Figure 76. Yellowstone Snowcoach – SNOVAN5, South Entrance (Jan 14th) Maximum Spectra for Low Speed and High Speed

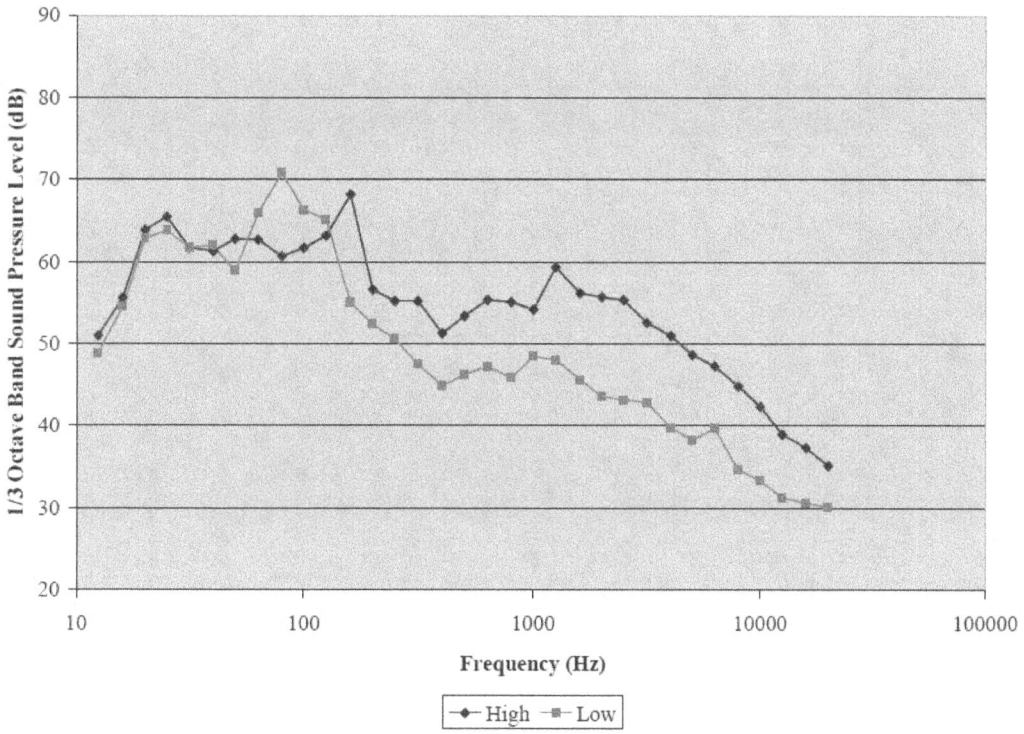

Figure 77. Alpen Guide – Kitty, South Entrance (Jan 14th) Maximum Spectra for Low Speed and High Speed

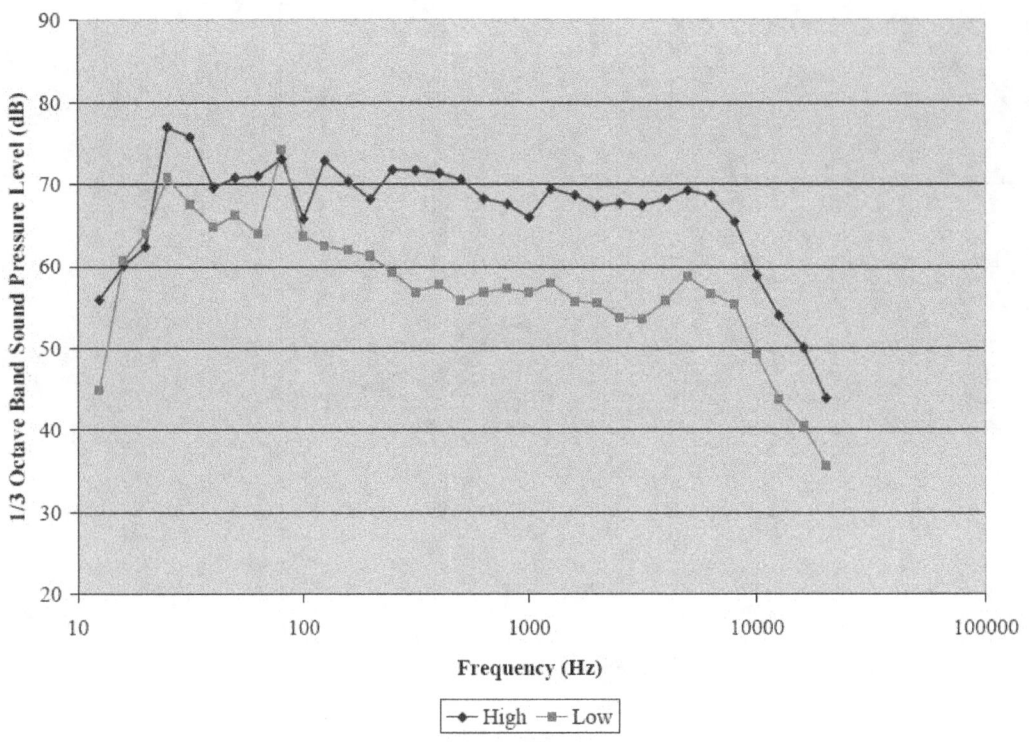

Figure 78. Xanterra 709, South Entrance (Jan 14th) Maximum Spectra for Low Speed and High Speed

Table 48. Maximum One-Third Octave Band Sound Levels for Select North Entrance Events (part 1), dB

Vehicle	Alpen Guide – Kitty	Alpen Guide – Kitty	Yellowstone Snowcoach – SNOVAN4	Yellowstone Snowcoach – SNOVAN4	Xanterra 713	Xanterra 713	Yellowstone Expedition – Hayden	Yellowstone Expedition – Hayden
Speed	High	Low	High	Low	High	Low	High	Low
Vehicle Side*	Right	Left	Right	Left	Right	Right	Left	Left
Max Time	12:23:13	12:44:50	13:31:04	13:58:15	14:48:19	15:23:46	16:17:39	16:35:06
One-Third Octave Band Center Frequency, (Hz)	Maximum Sound Pressure Level (dBA)							
12.5	64.7	70.2	69.8	72.5	69.1	63.0	73.0	57.6
16	64.4	66.0	69.7	73.4	68.9	66.9	71.2	53.1
20	68.0	68.8	68.0	73.3	69.5	68.4	73.2	58.8
25	66.8	68.6	66.0	70.8	73.0	62.4	71.3	58.8
31.5	66.2	64.5	64.2	67.6	65.3	69.1	65.9	57.2
40	64.2	62.1	65.0	61.7	62.8	65.8	62.5	55.7
50	65.4	61.2	66.8	61.4	69.7	63.1	63.4	62.3
63	63.6	70.2	64.5	61.5	62.7	69.8	65.5	67.7
80	62.2	69.2	58.5	68.1	68.0	71.5	68.9	61.2
100	61.7	61.4	56.6	57.7	65.2	59.9	64.7	59.0
125	65.0	59.4	57.8	52.4	71.2	60.9	57.9	59.8
160	66.0	54.4	67.7	59.0	61.5	61.7	66.3	63.1
200	58.5	53.5	69.8	56.1	67.0	66.9	62.0	59.7
250	61.0	53.2	65.4	59.9	70.6	61.2	60.7	57.1
315	60.4	50.3	74.1	61.1	61.8	55.3	57.0	53.9
400	56.9	49.3	72.0	58.0	65.2	56.2	56.5	53.1
500	57.8	50.2	69.6	54.6	64.8	55.6	57.3	54.2
630	59.4	49.7	67.1	54.8	71.6	54.9	56.1	52.0
800	58.0	48.5	66.8	54.6	66.6	55.7	54.6	50.6
1000	57.6	49.4	64.1	52.0	65.6	53.1	54.3	50.7
1250	57.9	49.4	60.7	49.2	64.4	52.4	51.0	47.1
1600	54.8	47.1	59.5	49.0	64.5	52.9	49.0	46.4
2000	54.3	45.0	59.7	50.1	63.1	50.7	49.5	46.3
2500	52.6	44.3	58.1	47.5	60.9	48.8	47.9	44.9
3150	51.3	41.0	56.8	45.6	60.0	47.2	47.1	45.6
4000	48.3	38.3	55.8	44.5	57.8	46.3	45.2	43.1
5000	46.2	37.3	54.7	42.7	57.5	47.9	44.4	41.5
6300	44.5	36.7	54.0	41.6	56.6	46.6	43.6	39.5
8000	42.7	33.3	52.5	40.3	54.6	43.7	42.1	37.5
10000	40.2	31.7	51.5	39.0	53.0	45.6	40.2	35.8
12500	37.3	29.7	51.0	37.8	52.0	43.7	38.2	34.2
16000	36.2	28.9	50.5	36.4	47.4	40.9	36.7	33.2
20000	34.9	28.1	49.0	34.5	44.2	42.4	35.1	31.8

* "Left/Right" indicates left/right side of vehicle from the driver's perspective.

Table 49. Maximum One-Third Octave Band Sound Levels for Select North Entrance Events (part 2), dB

Vehicle	Xanterra 707	Xanterra 707	Yellowstone Snowcoach – SNOVAN5	Yellowstone Snowcoach – SNOVAN5	Xanterra 537	Xanterra 537	Xanterra 430	Xanterra 430
Speed	High	Low	High	Low	High	Low	High	Low
Vehicle Side*	Left	Right	Right	Left	Left	Right	Left	Left
Max Time	11:25:45	12:09:39	13:02:43	13:15:08	14:30:47	15:15:55	15:43:09	16:15:56
One-Third Octave Band Center Frequency (Hz)	Maximum Sound Pressure Level (dBA)							
12.5	63.6	77.6	72.0	60.9	62.9	57.0	55.9	62.5
16	66.6	80.9	69.5	71.0	65.2	58.5	59.5	65.2
20	71.0	77.7	72.1	68.6	69.1	67.7	61.5	63.0
25	72.5	75.6	71.0	63.6	65.6	63.4	61.5	62.4
31.5	80.4	74.1	71.8	63.4	69.1	64.1	60.5	62.6
40	70.5	73.6	68.7	59.4	66.6	63.6	63.7	59.2
50	72.5	70.8	70.3	60.9	65.1	74.4	63.0	61.1
63	71.2	68.2	66.6	60.6	68.2	62.7	62.4	61.6
80	69.4	72.5	64.3	66.9	70.6	68.9	56.1	72.3
100	68.2	65.9	62.4	56.7	72.2	65.7	57.5	65.3
125	67.8	62.0	60.5	55.0	65.6	72.3	66.0	55.0
160	69.5	63.8	70.8	58.6	68.2	63.7	62.5	65.2
200	71.1	67.4	71.9	59.2	67.0	68.6	59.6	59.9
250	71.9	63.3	69.7	62.4	70.3	65.7	70.7	61.9
315	67.4	59.1	74.9	63.7	72.7	66.5	69.5	63.9
400	70.0	58.5	73.4	60.2	66.9	65.1	67.1	62.4
500	73.1	57.7	71.1	57.0	66.3	61.4	64.0	59.9
630	75.0	57.9	68.6	57.0	64.2	60.5	63.0	56.4
800	71.6	58.7	68.0	57.3	62.4	58.6	61.4	54.5
1000	70.3	57.8	66.9	56.7	64.7	57.2	58.5	50.7
1250	68.4	56.9	65.0	54.1	65.2	56.5	57.7	47.7
1600	68.6	57.1	63.8	53.6	66.4	60.7	55.2	46.9
2000	67.3	56.0	62.4	54.1	66.1	61.2	54.8	46.2
2500	66.2	54.3	60.4	53.2	64.1	59.5	54.5	45.1
3150	64.4	52.0	59.9	52.9	62.3	57.3	53.1	44.7
4000	62.8	49.6	57.6	49.8	61.8	58.2	51.4	44.2
5000	62.0	50.7	56.9	49.6	61.8	58.0	50.8	42.5
6300	60.7	49.0	55.7	47.4	56.7	48.1	49.3	40.6
8000	58.8	47.4	54.0	45.1	51.3	45.1	46.1	38.4
10000	56.9	46.8	53.7	43.4	46.8	38.7	44.8	37.4
12500	54.9	49.0	53.3	41.4	42.7	33.7	42.8	36.0
16000	53.1	45.8	53.4	40.0	42.4	31.8	40.8	35.0
20000	51.3	47.4	52.8	37.3	41.7	30.1	39.1	33.5

* "Left/Right" indicates left/right side of vehicle from the driver's perspective.

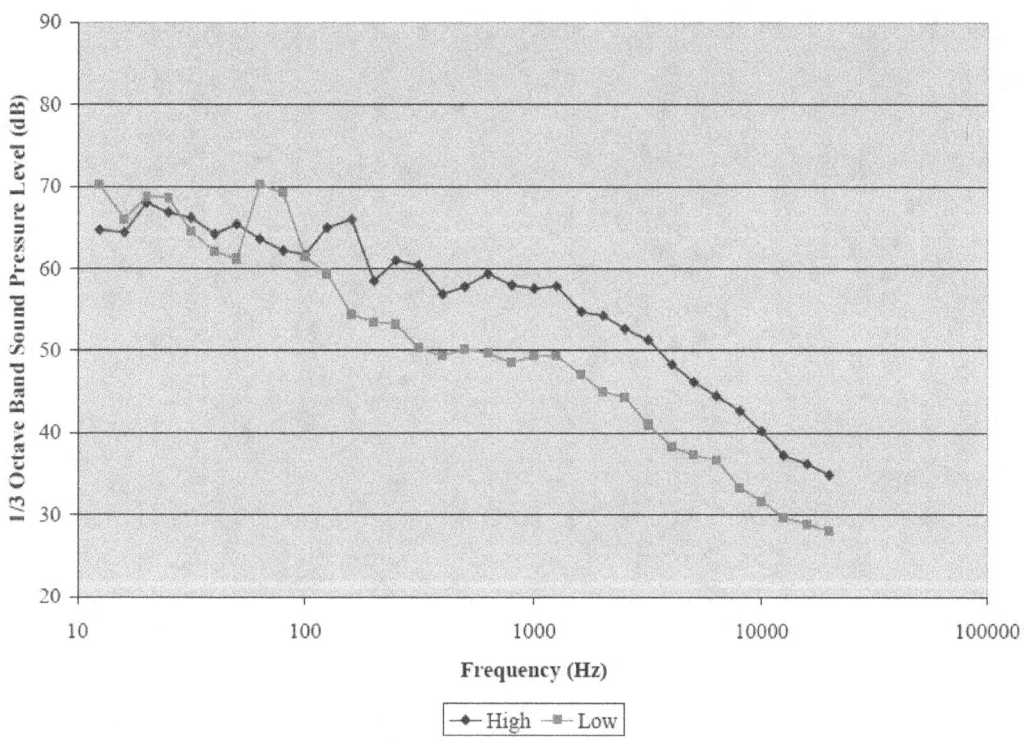

Figure 79. Alpen Guide – Kitty, North Entrance (Jan 15th) Maximum Spectra for Low Speed and High Speed

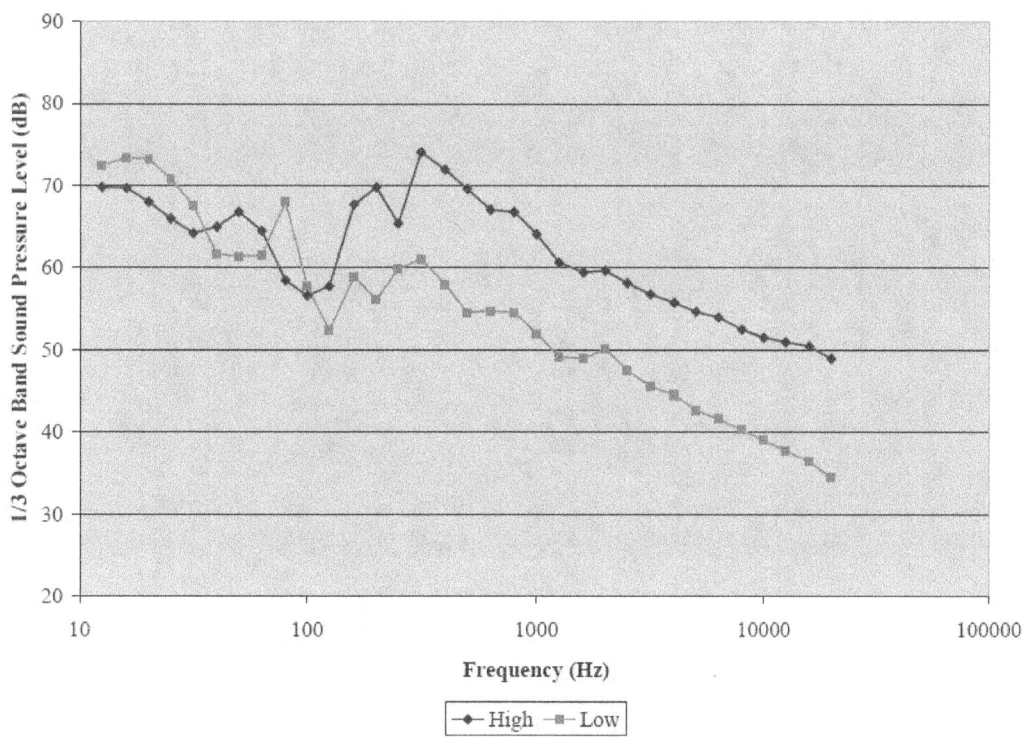

Figure 80. Yellowstone Snowcoach – SNOVAN4, North Entrance (Jan 15th) Maximum Spectra for Low Speed and High Speed

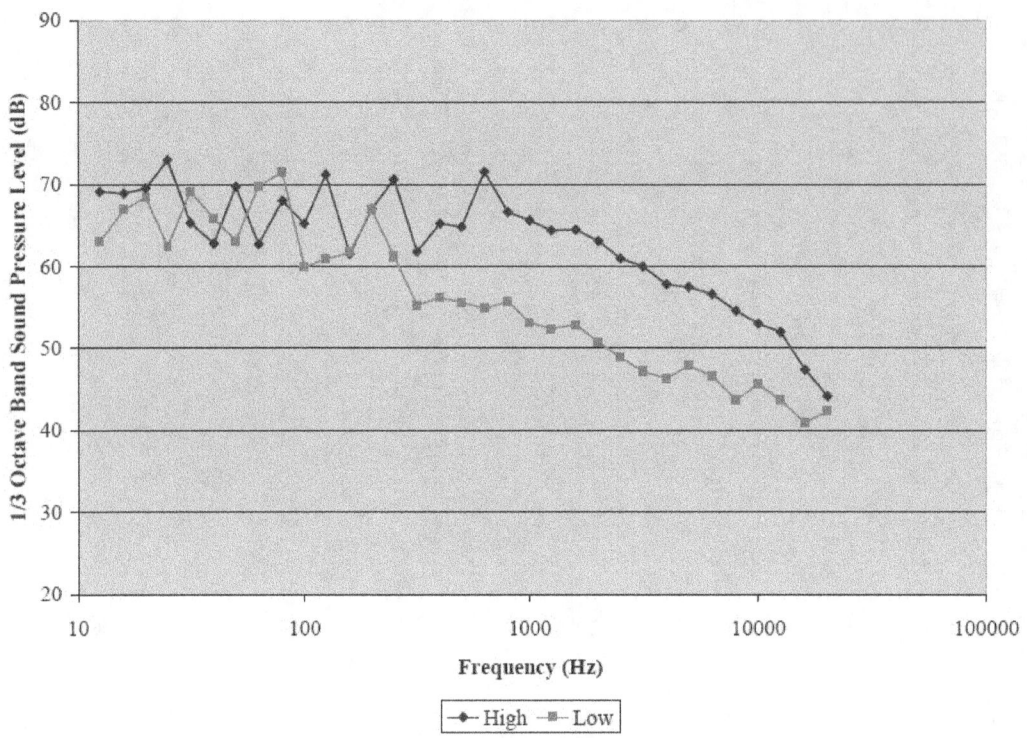

Figure 81. Xanterra 713, North Entrance (Jan 15th) Maximum Spectra for Low Speed and High Speed

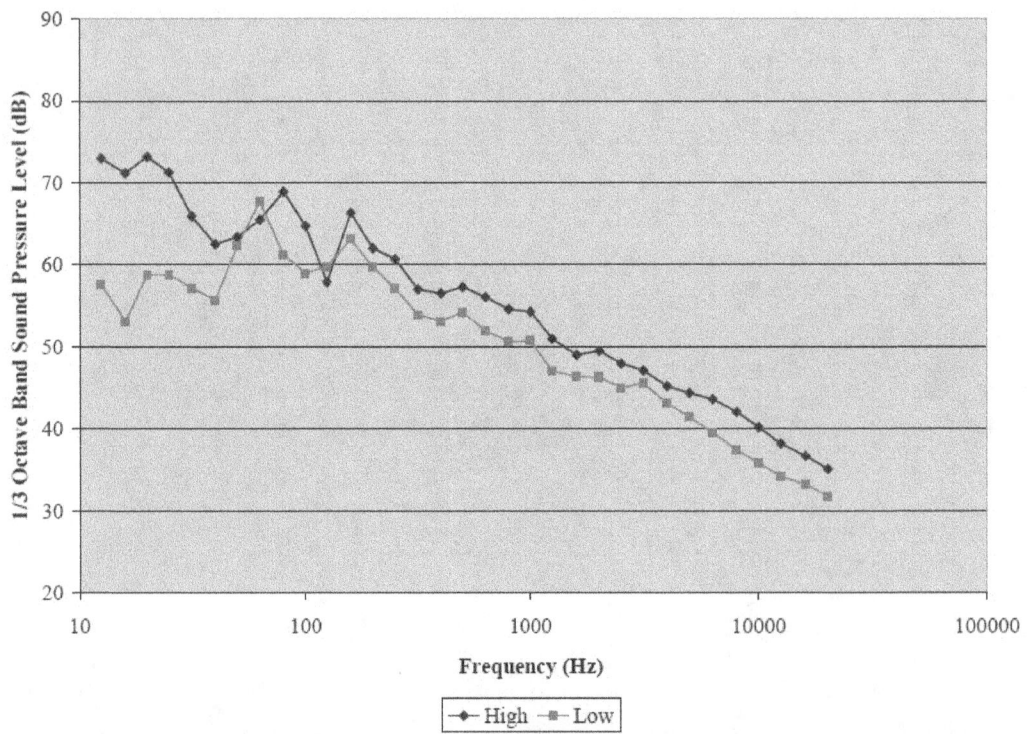

Figure 82. Yellowstone Expedition – Hayden, North Entrance (Jan 15th) Maximum Spectra for Low Speed and High Speed

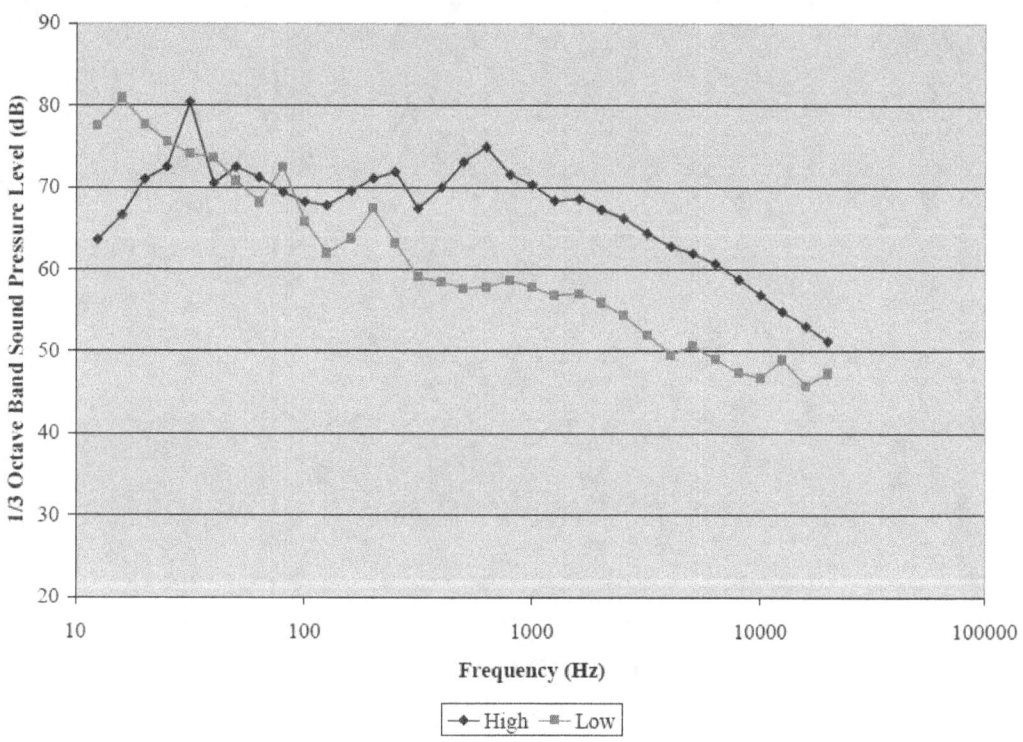

Figure 83. Xanterra 707, North Entrance (Jan 16th) Maximum Spectra for Low Speed and High Speed

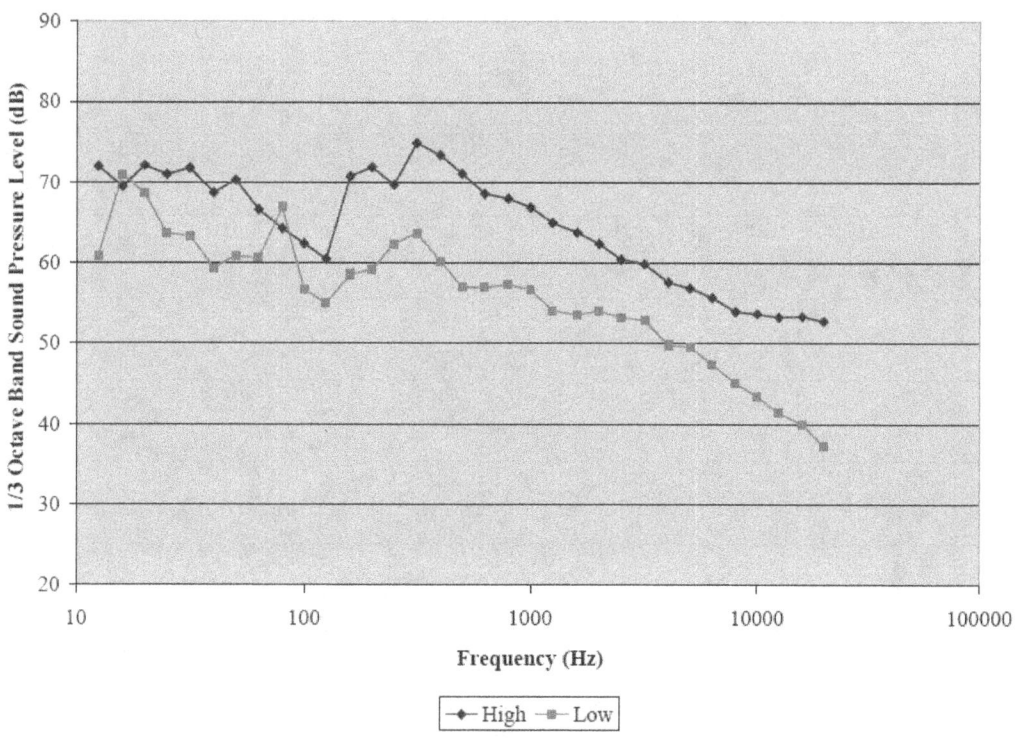

Figure 84. Yellowstone Snowcoach – SNOVAN5, North Entrance (Jan 16th) Maximum Spectra for Low Speed and High Speed

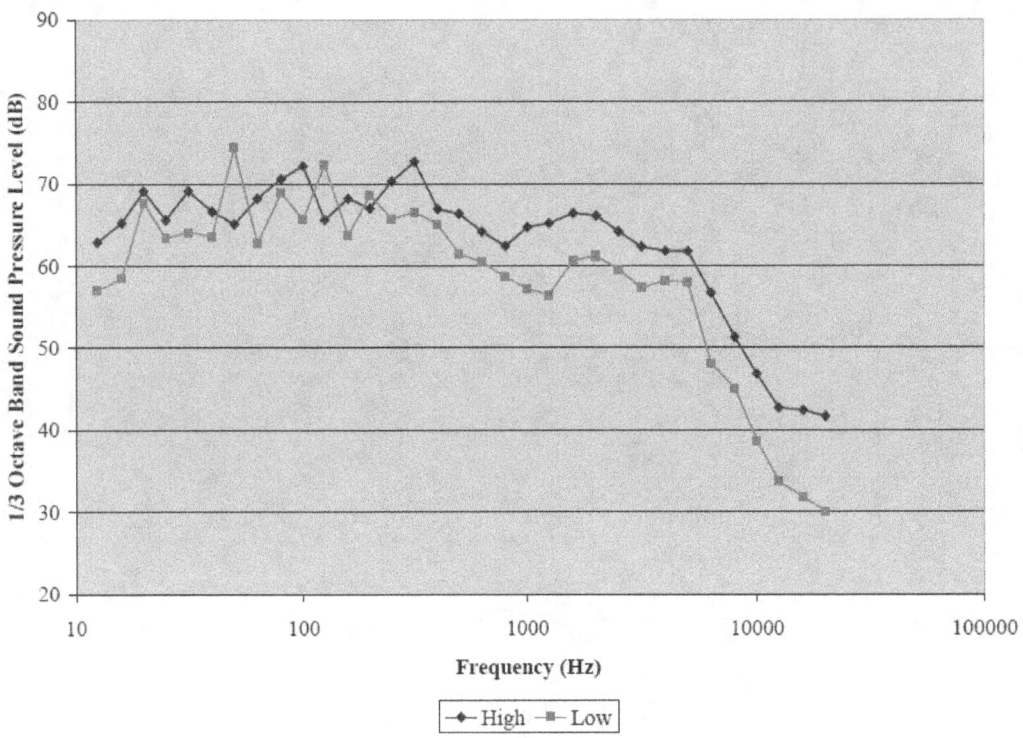

Figure 85. Xanterra 537 (Pernoth), North Entrance (Jan 16th) Maximum Spectra for Low Speed and High Speed

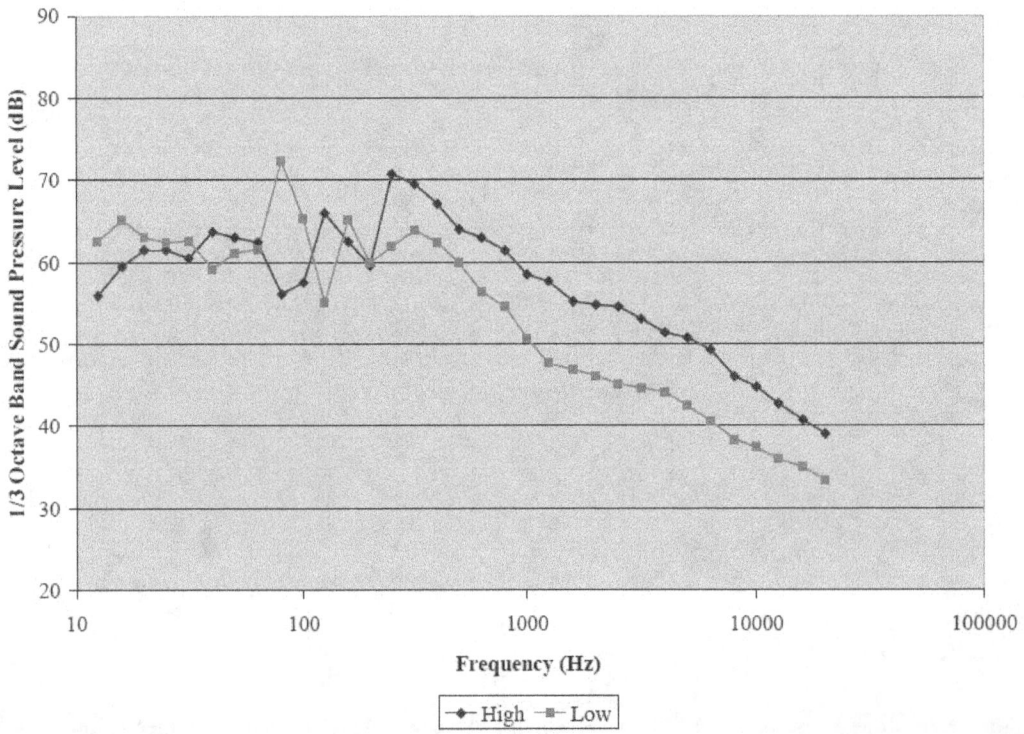

Figure 86. Xanterra 430, North Entrance (Jan 16th) Maximum Spectra for Low Speed and High Speed

Table 50. Maximum One-Third Octave Band Sound Levels for Select West Entrance Events (part 1), dB

Vehicle	Yellowstone Expedition – Hayden	Yellowstone Expedition – Hayden	Alpen Guide – Kitty	Alpen Guide – Kitty	Xanterra 713	Xanterra 713	Yellowstone Snowcoach – SNOVAN5	Yellowstone Snowcoach – SNOVAN5
Speed	High	Low	High	Low	High	Low	High	Low
Vehicle Side*	Left	Right	Right	Left	Right	Right	Left	Left
Max Time	12:24:53	12:48:39	13:40:18	13:36:26	14:20:03	14:45:27	15:27:09	15:13:05
One-Third Octave Band Center Frequency, (Hz)	Maximum Sound Pressure Level (dBA)							
12.5	51.7	54.9	52.5	58.8	51.0	48.2	60.7	45.9
16	48.0	54.8	53.8	56.8	58.6	57.7	64.4	65.7
20	52.0	59.9	66.3	60.7	66.8	64.7	71.7	63.9
25	58.6	56.0	61.7	62.8	67.5	65.5	70.9	58.5
31.5	57.3	55.5	64.8	63.2	75.0	69.4	64.7	60.0
40	56.8	53.9	62.2	59.6	64.6	62.9	65.7	55.5
50	64.0	55.9	62.0	61.5	64.0	60.0	65.3	53.6
63	62.9	68.6	62.2	64.6	67.2	68.6	66.7	54.0
80	62.4	64.8	58.2	70.0	62.9	73.0	61.7	65.1
100	58.9	60.8	61.3	64.1	65.7	60.1	60.5	51.1
125	69.3	59.7	61.3	54.8	64.1	62.8	61.5	50.6
160	68.1	59.9	70.5	58.3	72.5	64.4	69.2	59.5
200	64.1	59.6	61.2	56.4	64.8	66.4	73.2	55.8
250	66.6	59.0	60.9	56.7	68.5	63.3	70.4	63.5
315	66.8	60.3	63.0	53.9	66.2	59.0	73.6	64.7
400	66.6	59.4	58.4	51.8	66.7	59.3	74.8	62.1
500	64.7	60.1	59.4	51.2	71.2	55.7	72.2	57.1
630	62.8	56.7	58.5	50.4	66.8	56.8	68.3	55.5
800	59.7	54.3	57.2	50.0	67.9	57.4	67.5	56.0
1000	59.0	52.3	57.6	51.6	69.9	56.9	66.1	55.0
1250	57.5	53.1	56.5	52.3	67.0	53.2	63.7	52.0
1600	54.5	48.7	53.8	48.5	67.4	53.1	63.2	50.4
2000	54.6	47.9	53.1	47.0	64.7	51.4	63.7	50.8
2500	53.2	45.6	52.9	48.1	63.1	50.0	63.9	50.9
3150	52.1	44.8	51.3	44.9	62.2	48.7	63.8	51.8
4000	49.7	44.7	47.7	41.8	60.1	46.3	60.8	50.7
5000	49.5	47.2	45.8	40.3	59.3	47.6	60.3	51.8
6300	48.4	43.7	44.0	38.4	57.5	45.3	58.3	48.3
8000	46.8	39.8	42.1	36.8	56.5	44.0	56.2	45.5
10000	45.7	37.8	40.4	35.6	54.4	44.6	54.8	46.0
12500	44.3	36.4	38.6	33.3	52.2	42.2	51.9	42.6
16000	42.3	35.2	37.8	31.8	49.8	40.5	50.9	40.7
20000	40.5	33.3	36.9	30.2	48.7	39.9	48.1	37.3

* "Left/Right" indicates left/right side of vehicle from the driver's perspective.

Table 51. Maximum One-Third Octave Band Sound Levels for Select West Entrance Events (part 2), dB

Vehicle	See Yellowstone Tours #6	See Yellowstone Tours #6	Buffalo Bus Touring #4	Buffalo Bus Touring #4	Yellowstone Expedition – Eleanor	Yellowstone Expedition – Eleanor	Xanterra 710	Xanterra 710
Speed	High	Low	High	Low	High	Low	High	Low
Vehicle Side*	Right	Right	Left	Left	Left	Left	Right	Left
Max Time	10:44:43	11:12:03	12:12:16	12:29:34	13:33:08	13:59:16	14:33:37	15:00:12
One-Third Octave Band Center Frequency, (Hz)	Maximum Sound Pressure Level (dBA)							
12.5	59.8	54.5	54.6	60.5	53.0	60.8	53.0	48.8
16	59.6	62.4	56.4	65.6	57.7	59.5	61.2	55.2
20	65.1	77.0	64.7	75.3	66.8	55.6	63.0	60.1
25	67.6	75.1	67.0	70.4	63.3	60.8	67.0	63.0
31.5	62.5	74.7	66.6	66.7	61.3	55.2	73.5	64.8
40	66.8	65.7	68.1	68.1	65.5	57.1	66.2	64.9
50	64.2	66.3	63.2	77.5	63.0	61.3	69.3	62.5
63	64.8	59.7	67.8	58.6	64.6	63.5	68.2	65.6
80	57.8	72.6	63.2	56.6	60.0	65.5	68.7	64.8
100	57.4	58.7	58.2	59.9	63.0	58.5	65.4	62.5
125	58.2	55.7	59.3	56.3	64.4	60.1	67.0	62.1
160	65.8	63.7	60.8	56.7	62.6	60.6	70.1	61.1
200	76.2	61.0	67.3	61.3	64.0	58.2	68.5	63.7
250	69.9	64.8	63.8	59.5	63.0	58.3	72.3	60.4
315	73.8	65.6	63.4	59.5	63.4	58.2	67.1	57.4
400	76.6	63.6	60.6	56.6	63.1	58.7	68.1	55.9
500	71.8	61.2	59.2	54.1	62.1	56.8	65.2	54.7
630	69.3	58.2	59.8	54.0	60.8	55.1	67.7	54.3
800	68.8	58.8	61.8	53.9	59.2	54.0	67.5	54.4
1000	67.1	58.2	61.8	53.4	60.4	55.4	67.7	54.4
1250	65.3	58.9	61.5	53.1	60.0	56.8	66.3	51.4
1600	64.1	57.0	59.9	53.9	57.6	55.0	65.3	51.7
2000	62.3	55.0	59.6	53.3	55.4	51.9	64.1	50.2
2500	61.1	54.3	57.3	50.9	54.0	50.4	62.0	50.4
3150	60.7	56.9	56.9	51.5	51.8	48.0	60.3	49.6
4000	59.3	54.5	57.7	51.7	48.7	44.3	58.6	48.5
5000	58.0	54.6	57.4	51.3	48.3	44.6	57.9	50.9
6300	57.3	53.5	55.7	49.5	46.4	40.8	56.4	49.5
8000	55.3	54.4	56.1	48.1	44.4	41.0	55.3	43.7
10000	53.7	53.4	55.5	47.5	42.8	38.3	53.3	44.7
12500	52.0	48.8	54.4	46.8	40.8	36.4	51.0	42.6
16000	51.1	46.3	52.6	47.7	39.3	35.5	49.7	41.4
20000	48.8	43.2	51.0	45.2	39.0	35.9	50.2	43.6

* "Left/Right" indicates left/right side of vehicle from the driver's perspective.

Table 52. Maximum One-Third Octave Band Sound Levels for Select West Entrance Events (part 3), dB

Vehicle		Buffalo Bus Touring T2	Buffalo Bus Touring T2	See Yellowstone Tours #9	See Yellowstone Tours #9	Buffalo Bus Touring #3	Buffalo Bus Touring #3	See Yellowstone Tours #4	See Yellowstone Tours #4
Speed		High	Low	High	Low	High	Low	High	Low
Vehicle Side*		Left	Left	Left	Right	Left	Left	Right	N/A
Max Time		15:32:57	15:56:39	10:38:08	11:10:32	11:48:34	12:13:00	12:43:44	N/A
		Maximum Sound Pressure Level (dBA)							
One-Third Octave Band Center Frequency, (Hz)	12.5	52.8	56.2	63.2	60.5	49.3	52.3	61.4	N/A
	16	58.9	61.2	69.2	62.7	56.5	57.4	61.0	N/A
	20	63.7	60.4	67.0	64.2	64.6	66.1	58.3	N/A
	25	66.6	69.6	66.6	70.2	65.3	71.1	66.2	N/A
	31.5	64.7	79.7	67.4	70.5	66.3	70.2	59.9	N/A
	40	64.4	68.6	66.7	66.1	61.2	64.7	59.7	N/A
	50	68.2	63.4	64.8	58.7	68.0	61.5	63.5	N/A
	63	58.7	66.6	60.0	69.1	61.9	64.7	61.9	N/A
	80	55.6	63.4	60.5	57.4	53.9	60.1	55.9	N/A
	100	64.6	56.0	58.0	58.1	59.5	55.6	62.7	N/A
	125	68.9	55.9	60.5	59.9	68.7	57.2	56.1	N/A
	160	60.8	63.1	76.0	59.9	61.9	59.8	68.5	N/A
	200	62.7	58.1	72.0	64.1	62.6	58.2	65.0	N/A
	250	71.6	63.1	67.5	65.5	75.0	61.8	62.4	N/A
	315	69.2	65.0	75.6	67.0	70.9	64.7	73.6	N/A
	400	71.5	63.3	71.9	67.3	73.3	63.3	68.7	N/A
	500	66.3	60.2	75.5	66.0	68.9	60.1	67.5	N/A
	630	63.2	57.0	69.6	66.9	65.3	58.4	63.9	N/A
	800	63.6	57.2	69.2	71.4	63.7	56.8	62.6	N/A
	1000	60.5	55.2	67.5	67.5	61.5	54.3	60.3	N/A
	1250	57.2	52.7	63.8	63.9	58.6	52.9	55.8	N/A
	1600	54.6	51.4	61.0	61.1	56.6	52.3	53.6	N/A
	2000	53.6	49.9	60.4	59.6	56.0	51.7	52.4	N/A
	2500	54.3	50.2	59.9	58.7	56.9	50.3	52.6	N/A
	3150	53.9	49.2	59.0	58.2	56.1	50.2	52.2	N/A
	4000	52.1	48.3	57.1	52.7	53.9	48.3	50.3	N/A
	5000	51.9	46.5	56.5	51.7	54.4	47.7	49.5	N/A
	6300	51.0	46.5	54.5	49.1	52.6	45.7	49.1	N/A
	8000	49.1	44.0	53.8	48.2	50.1	43.2	48.1	N/A
	10000	47.0	43.3	51.0	46.0	48.1	42.5	48.3	N/A
	12500	45.7	41.0	48.5	42.7	46.3	40.8	45.5	N/A
	16000	44.1	39.5	47.7	38.1	44.8	39.2	44.2	N/A
	20000	42.9	38.1	45.4	35.1	42.4	36.5	41.8	N/A

* "Left/Right" indicates left/right side of vehicle from the driver's perspective.

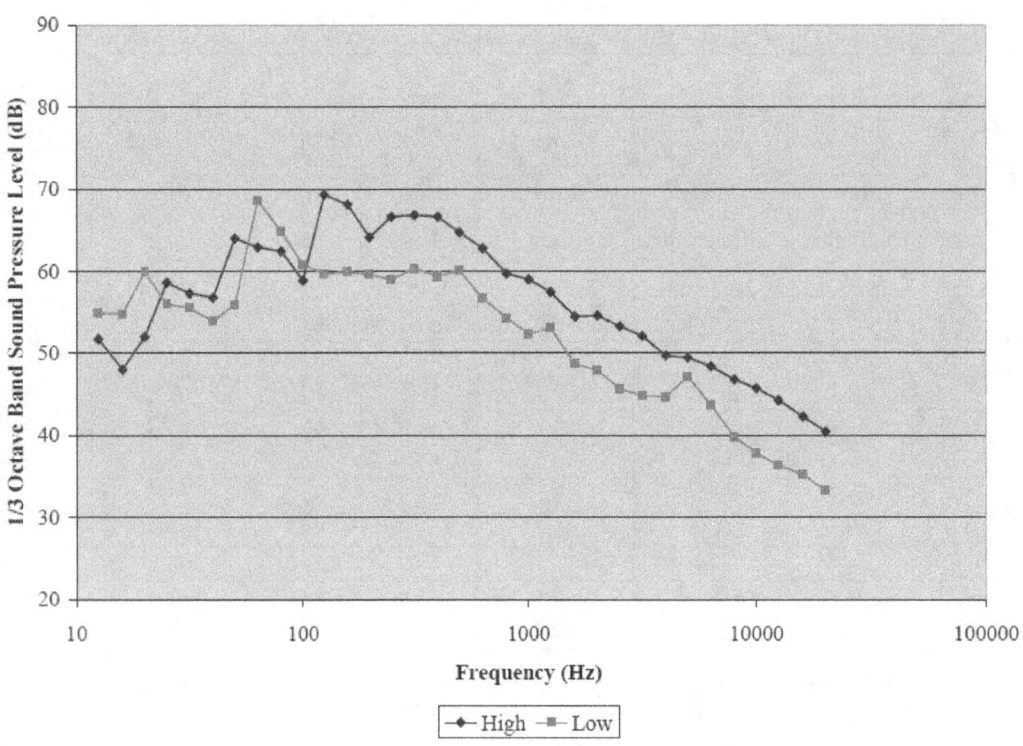

Figure 87. Yellowstone Expedition – Hayden, West Entrance (Jan 20th) Maximum Spectra for Low Speed and High Speed

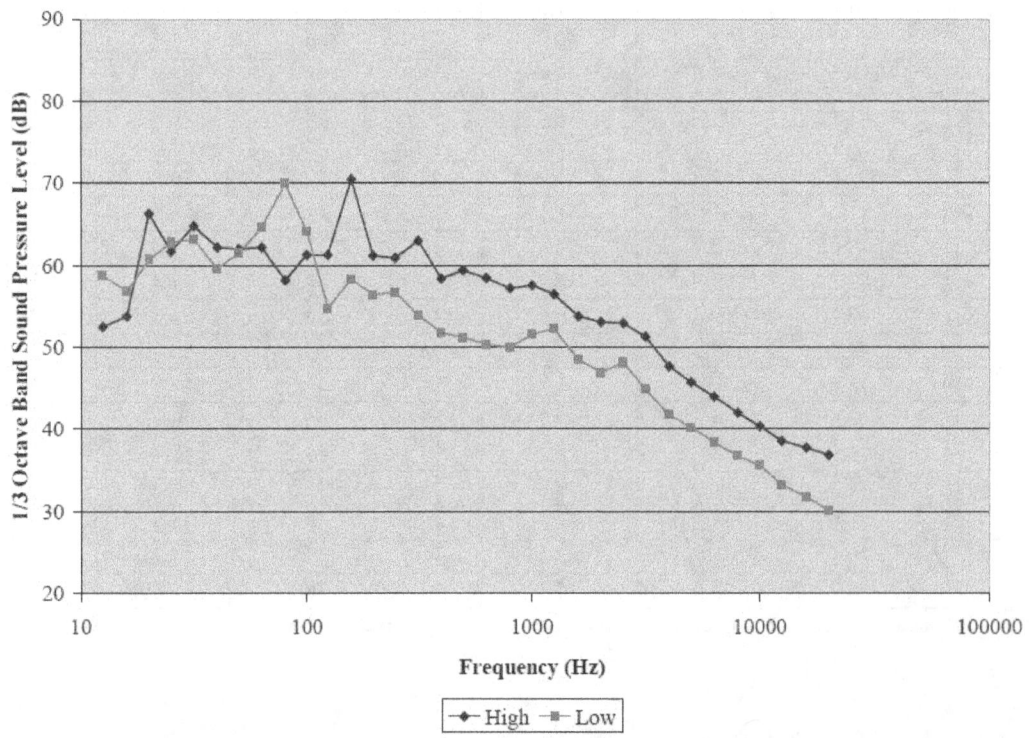

Figure 88. Alpen Guide – Kitty, West Entrance (Jan 20th) Maximum Spectra for Low Speed and High Speed

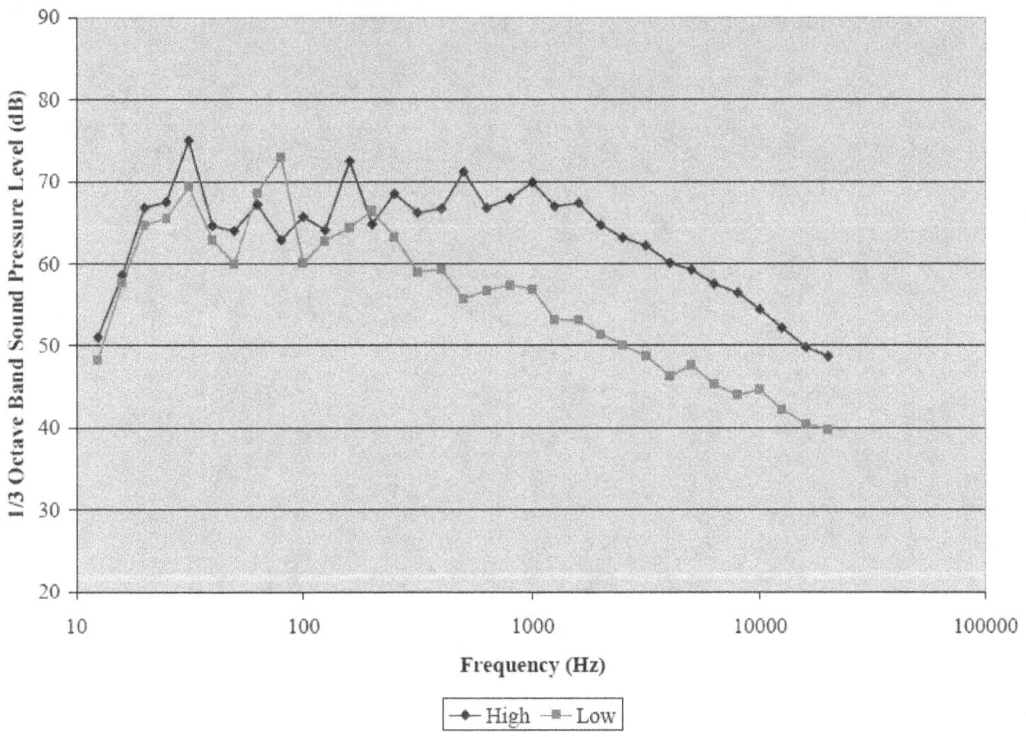

Figure 89. Xanterra 713, West Entrance (Jan 20th) Maximum Spectra for Low Speed and High Speed

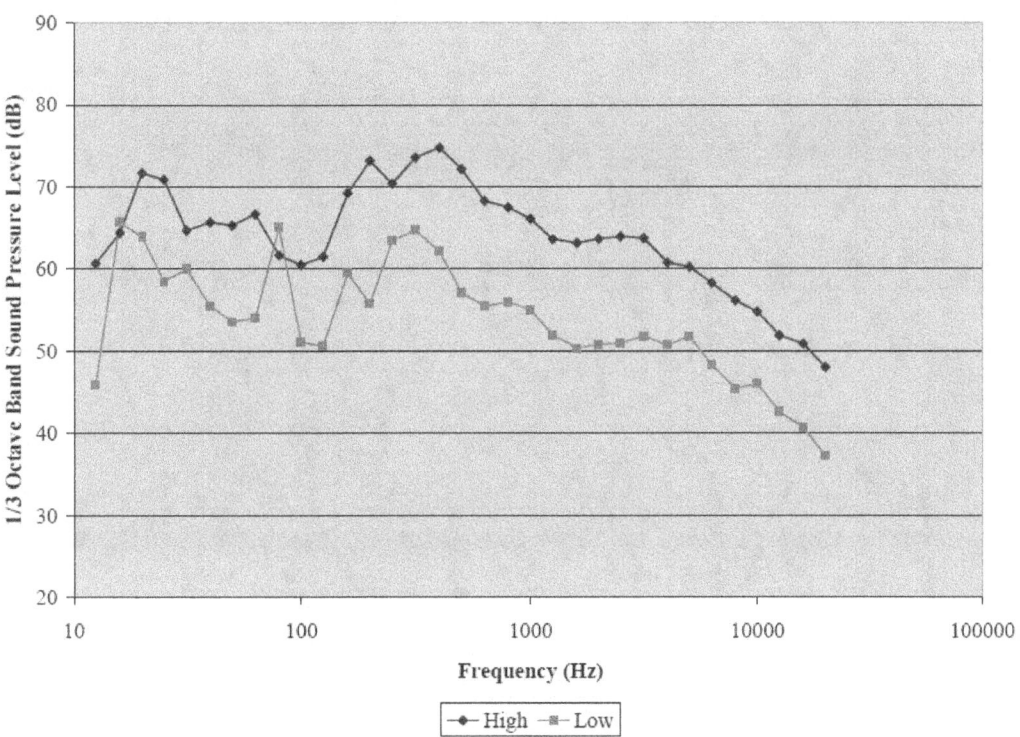

Figure 90. Yellowstone Snowcoach – SNOVAN5, West Entrance (Jan 20th) Maximum Spectra for Low Speed and High Speed

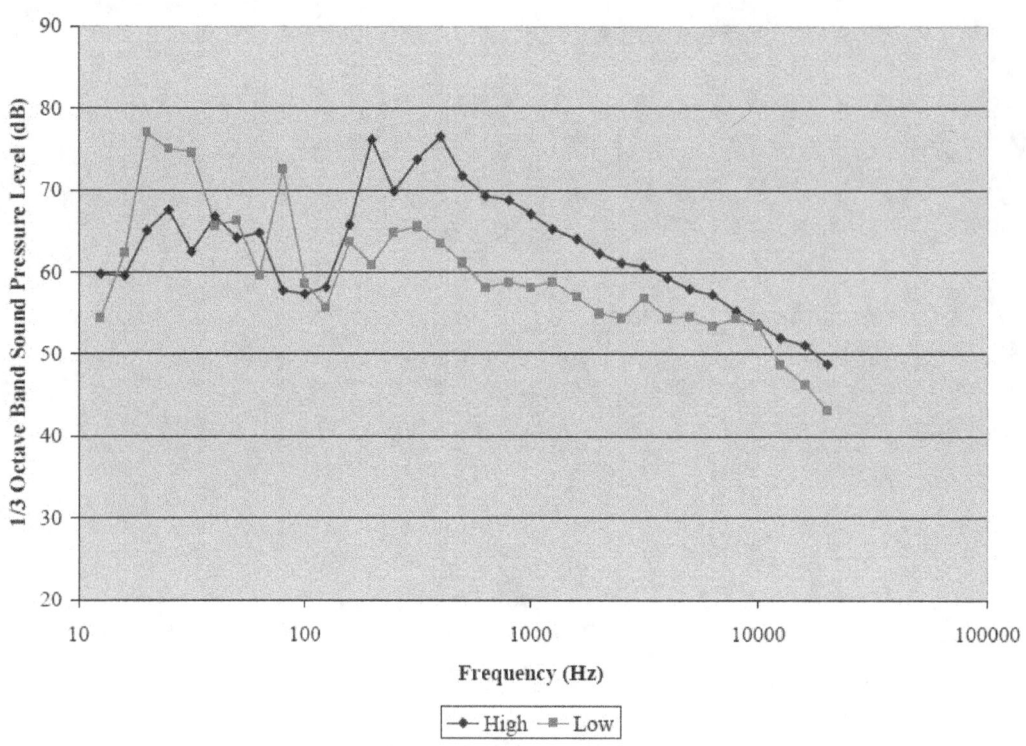

Figure 91. See Yellowstone Tours #6, West Entrance (Jan 21st) Maximum Spectra for Low Speed and High Speed

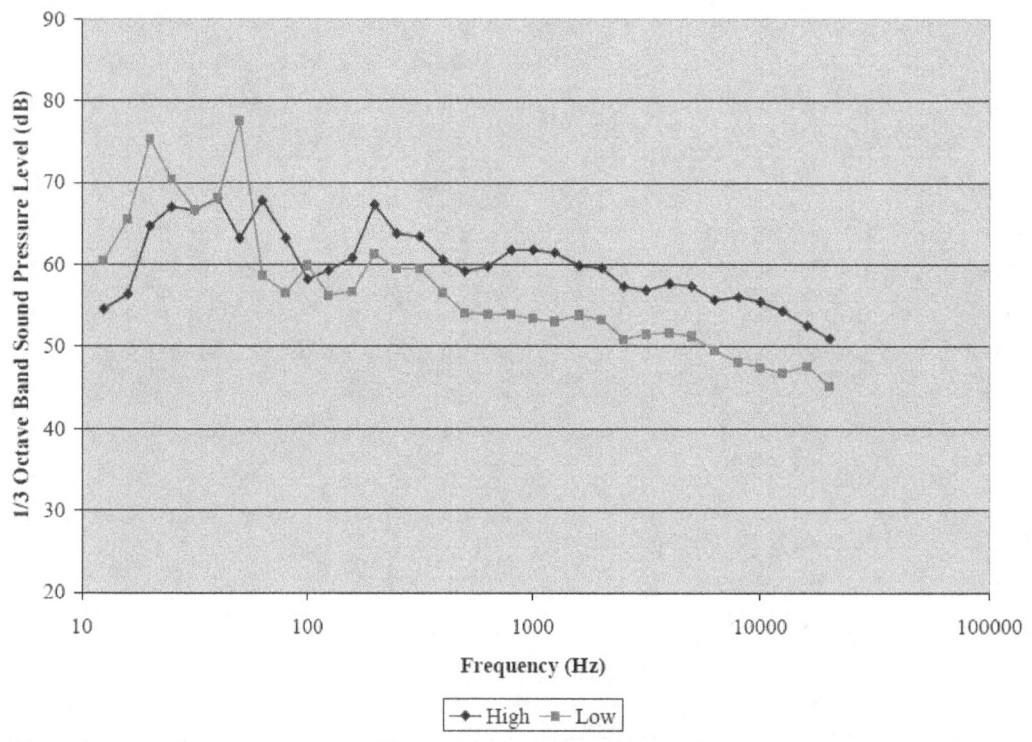

Figure 92. Buffalo Bus Touring #4, West Entrance (Jan 21st) Maximum Spectra for Low Speed and High Speed

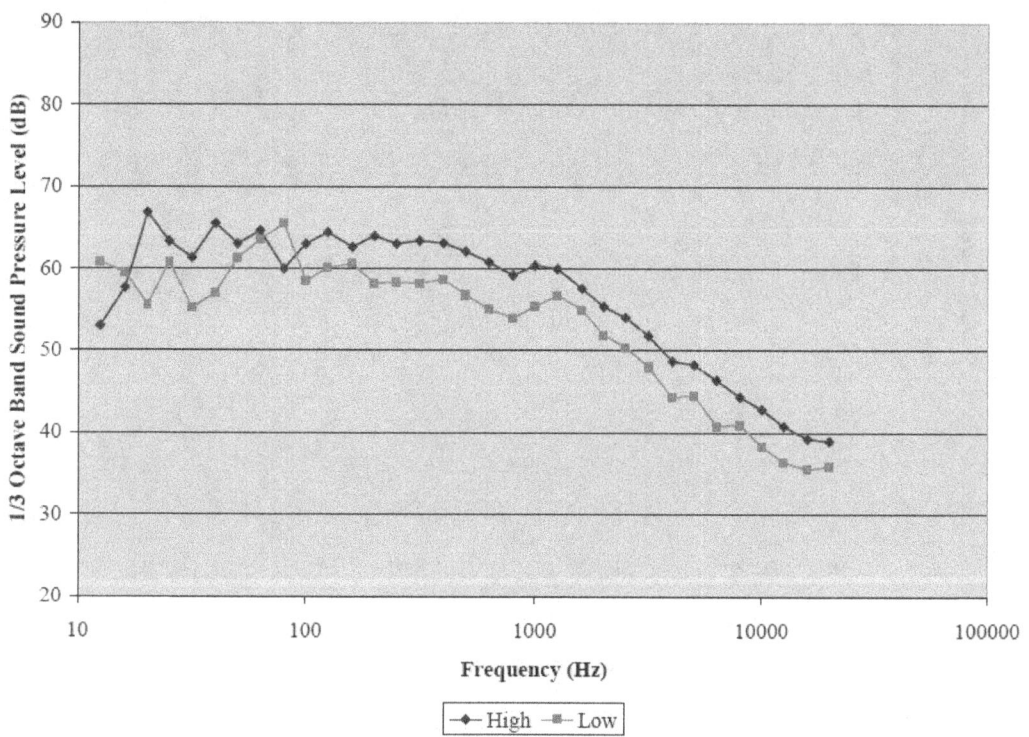

Figure 93. Yellowstone Expedition – Eleanor, West Entrance (Jan 21st) Maximum Spectra for Low Speed and High Speed

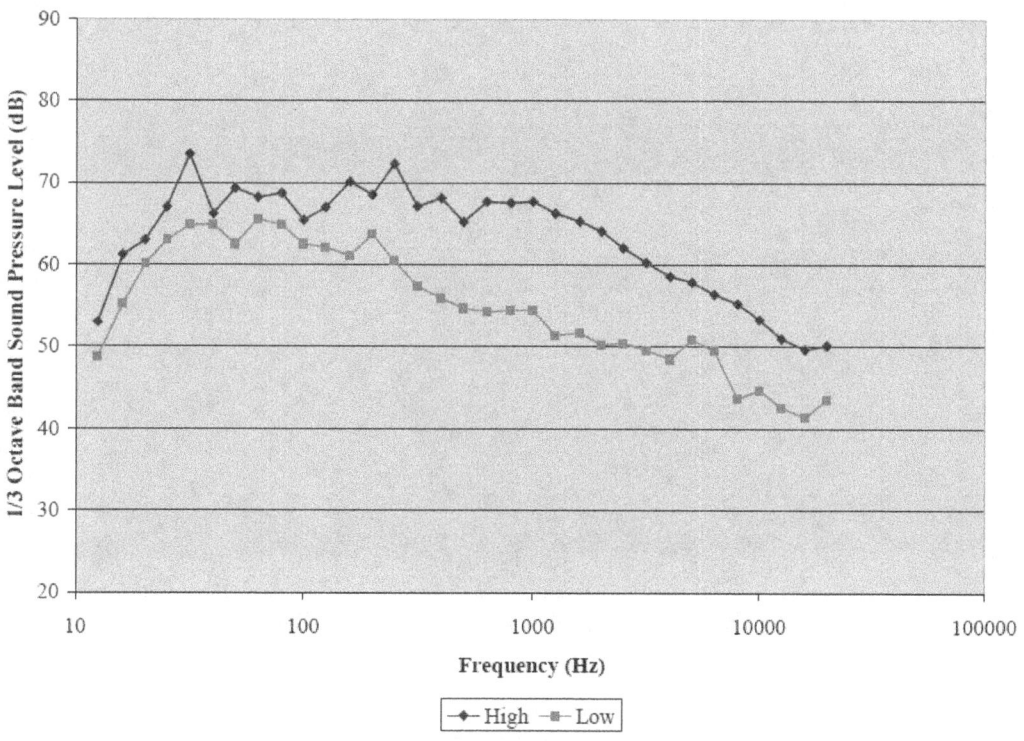

Figure 94. Xanterra 710, West Entrance (Jan 21st) Maximum Spectra for Low Speed and High Speed

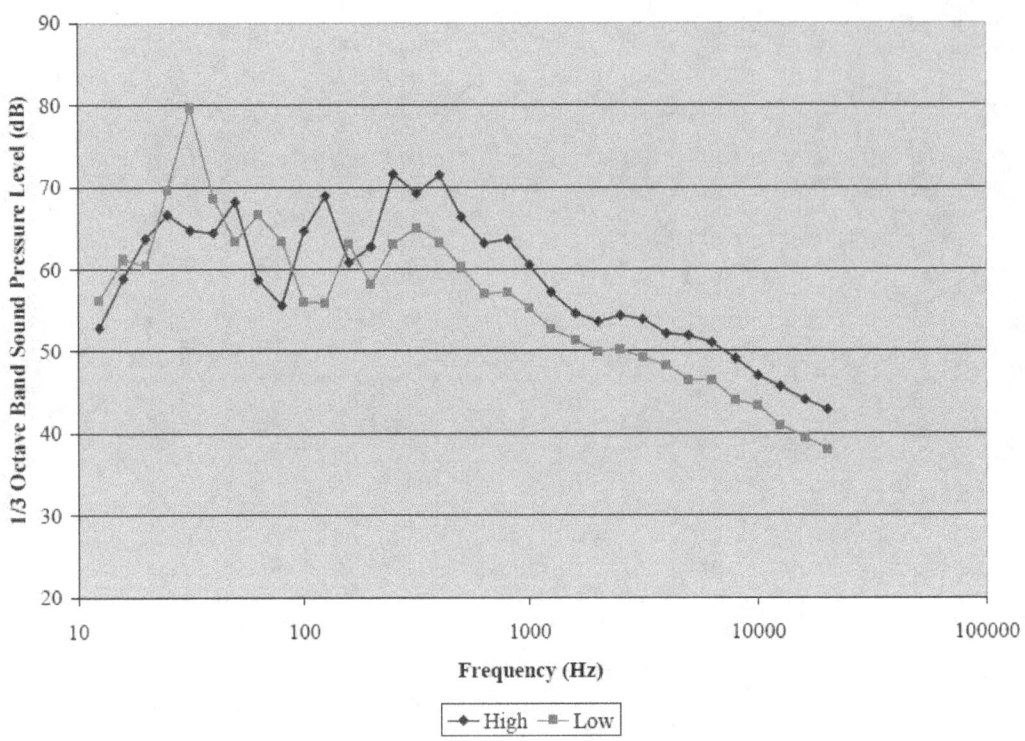

Figure 95. Buffalo Bus Touring T2, West Entrance (Jan 21st) Maximum Spectra for Low Speed and High Speed

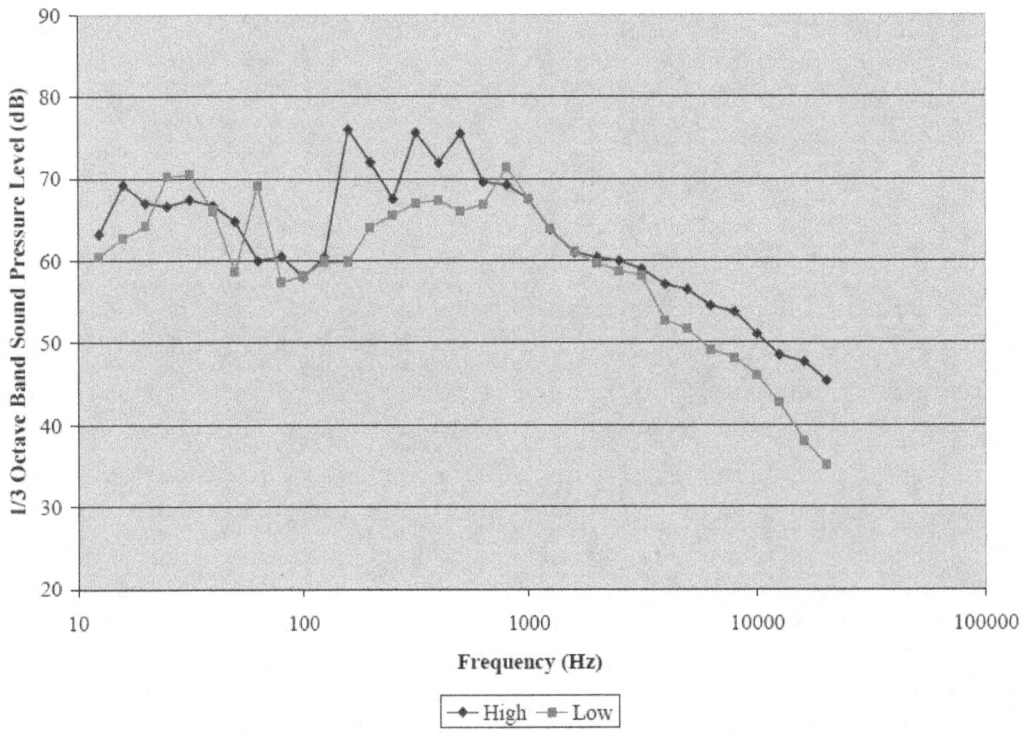

Figure 96. See Yellowstone Tours #9, West Entrance (Jan 22nd) Maximum Spectra for Low Speed and High Speed

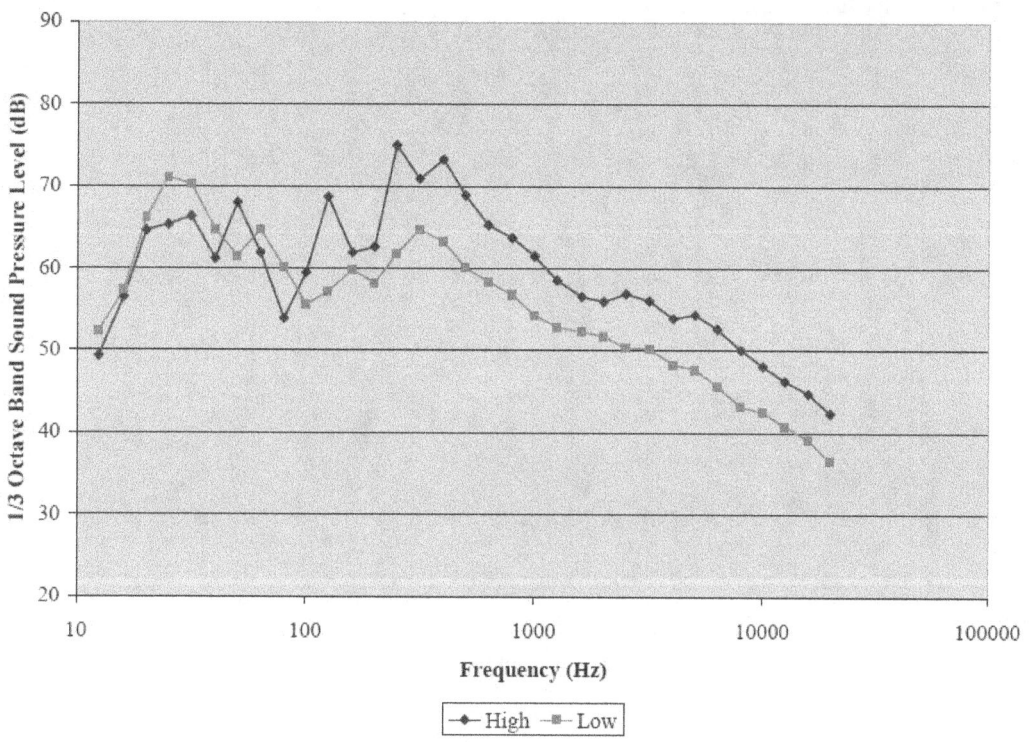

Figure 97. Buffalo Bus Touring #3, West Entrance (Jan 22nd) Maximum Spectra for Low Speed and High Speed

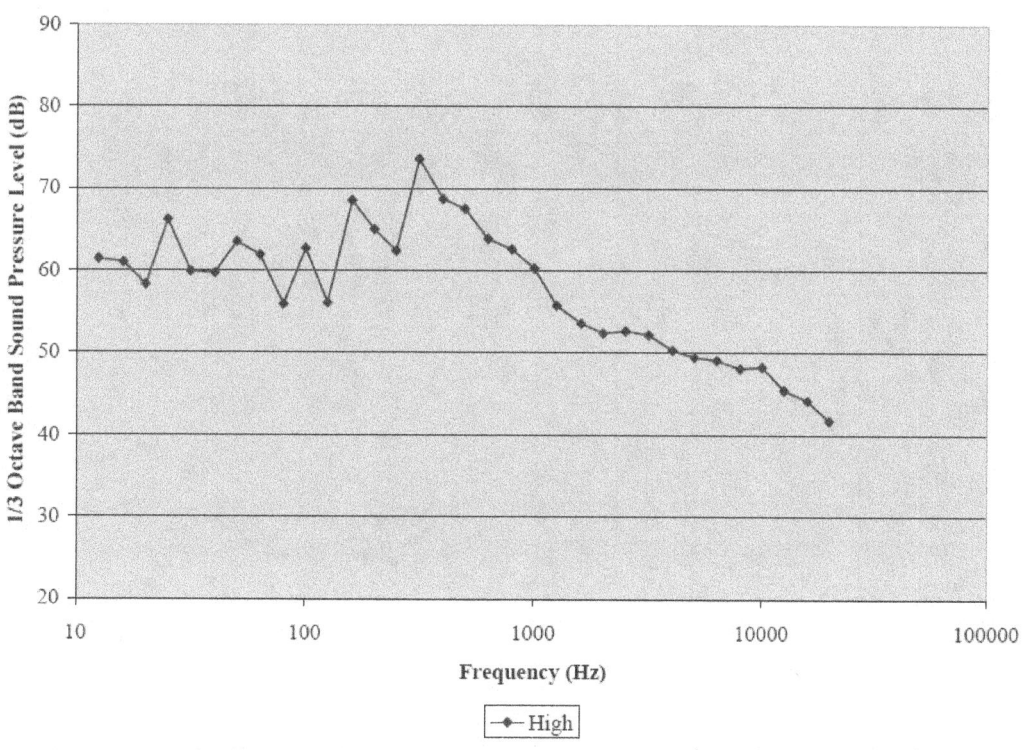

Figure 98. See Yellowstone Tours #4, West Entrance (Jan 22nd) Maximum Spectra for High Speed

This page intentionally left blank

Appendix G: Measurement Site Bias Data

This appendix presents an analysis of the data comparing the four snowcoaches tested at each of the three sites. This includes numerical values for L_{ASmx} and time history plots for each vehicle comparing each type of event.

Table 53 and Table 55 show a comparison of the four coaches tested at each site for low speeds and high speeds and idle, respectively.

Table 53. Low Speed L_{ASmx} for Snowcoaches at All Sites

Vehicle	Entrance	Vehicle Side*	Average L_{ASmx} (dBA)	Average Speed of Runs (mph)
Xanterra 713	South	Right	66	16
	West	Right	66	15
	North	Right	65	15
Alpen Guide – Kitty	West	Left	61	16
	North	Left**	59	15
	South	Right**	58	15
Yellowstone Snowcoach – SNOVAN5	West	Left	66	15
	North	Left	66	16
	South	Right	59	15
Yellowstone Expedition – Hayden	West	Right	64	16
	North	Left	60	15
	South	Left	59	15

* "Left/Right" indicates left/right side of vehicle from the driver's perspective.
** Indicates that data was only available for one side of the vehicle.

Table 54. High Speed L_{ASmx} for Snowcoaches at All Sites

Vehicle	Entrance	Vehicle Side[*]	Average L_{ASmx} (dBA)	Average Speed of Runs (mph)
Xanterra 713	West	Right	77	32
	North	Right	76	26
	South	Left	75	23
Alpen Guide – Kitty	West	Right	67	32
	North	Right	67	29
	South	Right[**]	66	32
Yellowstone Snowcoach – SNOVAN5	West	Left	78	36
	North	Right	77	34
	South	Right	68	25
Yellowstone Expedition – Hayden	West	Left	70	30
	North	Left	64	20
	South	Left	64	20[***]

[*] "Left/Right" indicates left/right side of vehicle from the driver's perspective.
[**] Indicates that data was only available for one side of the vehicle.
[***] Due to conditions this was maximum safe speed.

Table 55. Idle L_{Aeq} for Snowcoaches at All Sites

Vehicle	Entrance	Vehicle Side[*]	Average L_{ASmx} (dBA)	Daily Maximum Ambient Sound Level (dBA)	
				Minimum	Maximum
Xanterra 713	West	Left	46	22	24
	North	Left	45[**]	32	42
	South	Left	43	22	24
Alpen Guide – Kitty	North	Right	41[**]	32	42
	South	Right	39	22	24
	West	Right	38	22	24
Yellowstone Snowcoach – SNOVAN5	North	Right	41	28	34
	West	Left	40	22	24
	South	Right	38	22	24
Yellowstone Expedition – Hayden	West	Left	40	22	24
	North	Left	41[**]	32	42
	South	Left	39	22	24

[*] "Left/Right" indicates left/right side of vehicle from the driver's perspective.
[**] Idle measurements with sound levels that are within 10 dB of the ambient sound level

Figure 99 though Figure 110 show the time history comparisons for low speed and high speed as well as idle measurements for each of the four snowcoaches. It should be noted that due to different ambient sound levels at each site, some of the time histories start off with higher a sound level; this is most noticeable at the north entrance site. The ambient at the north entrance was as high as 42 dBA, mostly due to higher wind speeds; all of the events represented were within tolerance, less than 12 mph.

Examining the Alpen Guide – Kitty in Figure 99 though Figure 101, it can be seen that both the low and high speed events line up closely between the three sites. The high ambient sound levels are likely the cause for the north entrance having the loudest idle measurements.

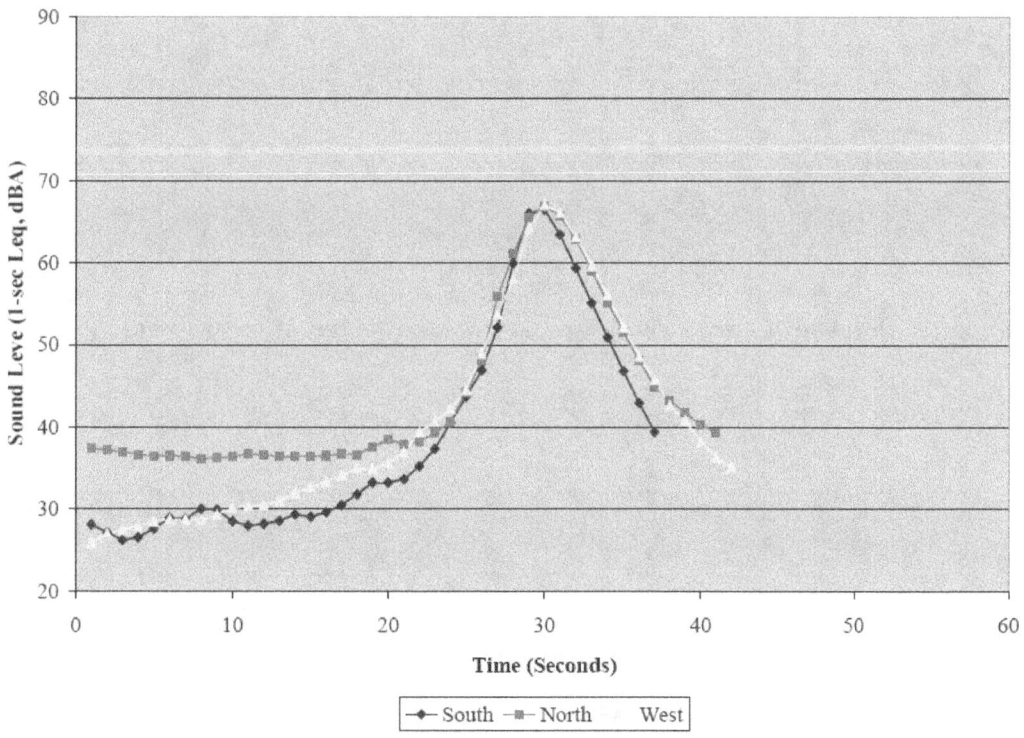

Figure 99. Alpen Guide – Kitty High Speed Time History Comparison

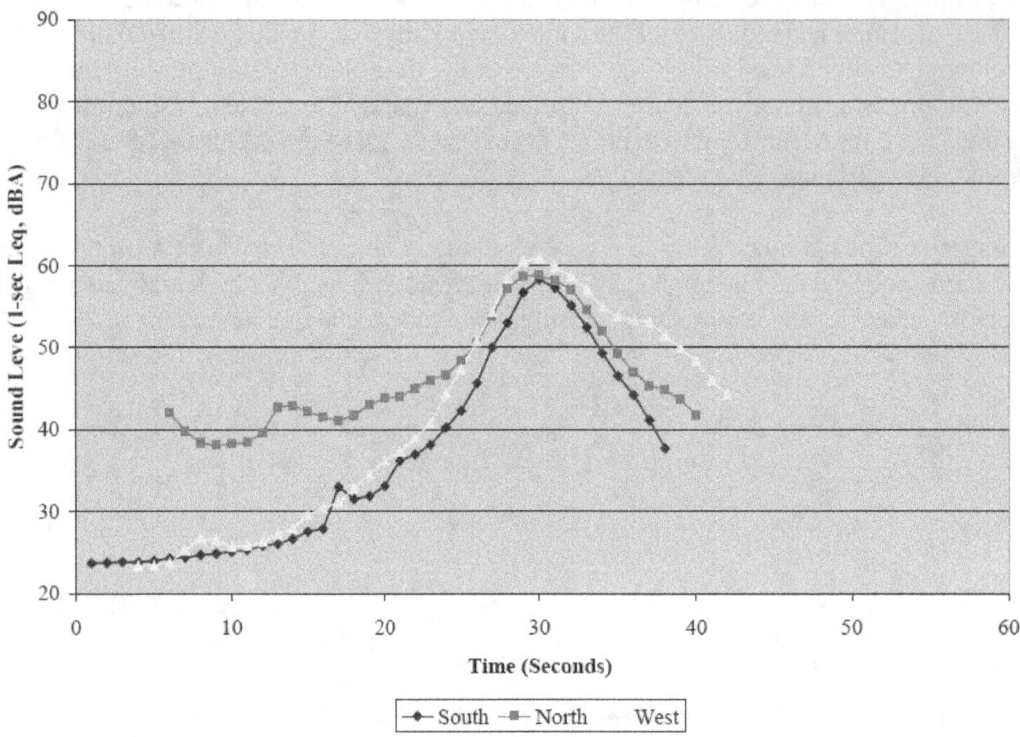

Figure 100. Alpen Guide – Kitty, Low Speed Time History Comparison

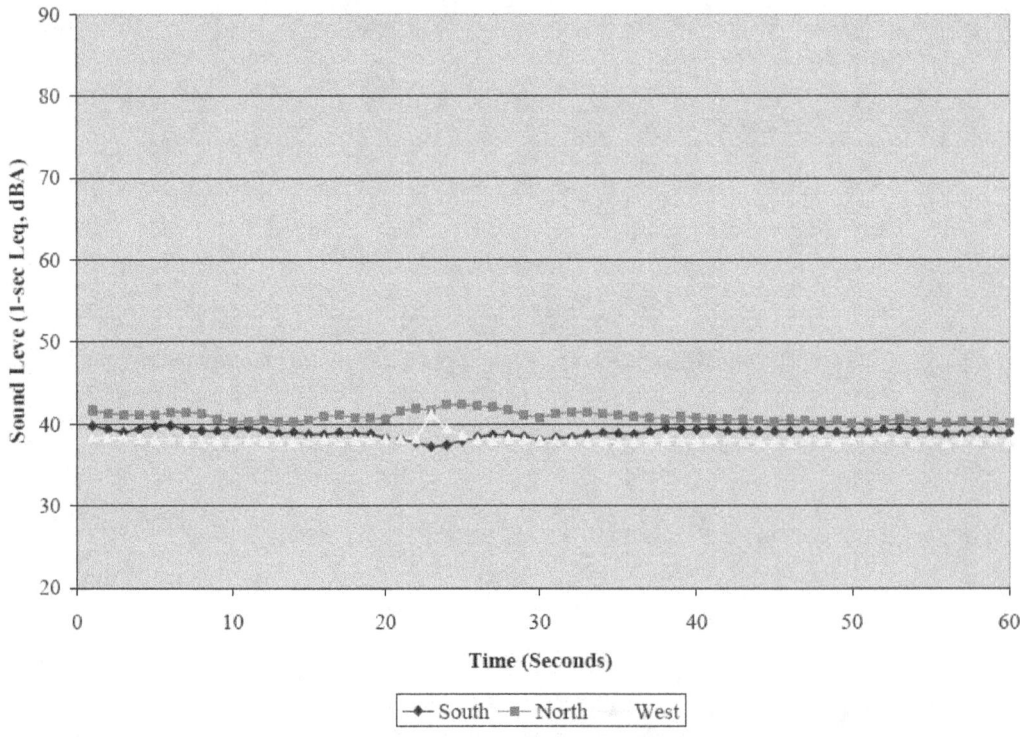

Figure 101. Alpen Guide – Kitty, Idle Time History Comparison

Examining the Yellowstone Snowcoach – SNOVAN5 in Figure 102 though Figure 104, it should be noted that the north entrance had much louder ambient sound levels, as seen by the start of the events. The peaks of the north and west have good agreement while the south entrance measurements are much lower in sound level. The high speed event at the south entrance was 8 mph slower than the north entrance and 10 mph slower than the west entrance which likely contributed to the difference in sound level. Speed difference alone cannot account for all site-to-site differences as can be seen by the low speed data comparison, where speed differences were small, in these cases the effect of the snow berm should also be considered. The snow berm is discussed in section 4.3.2.

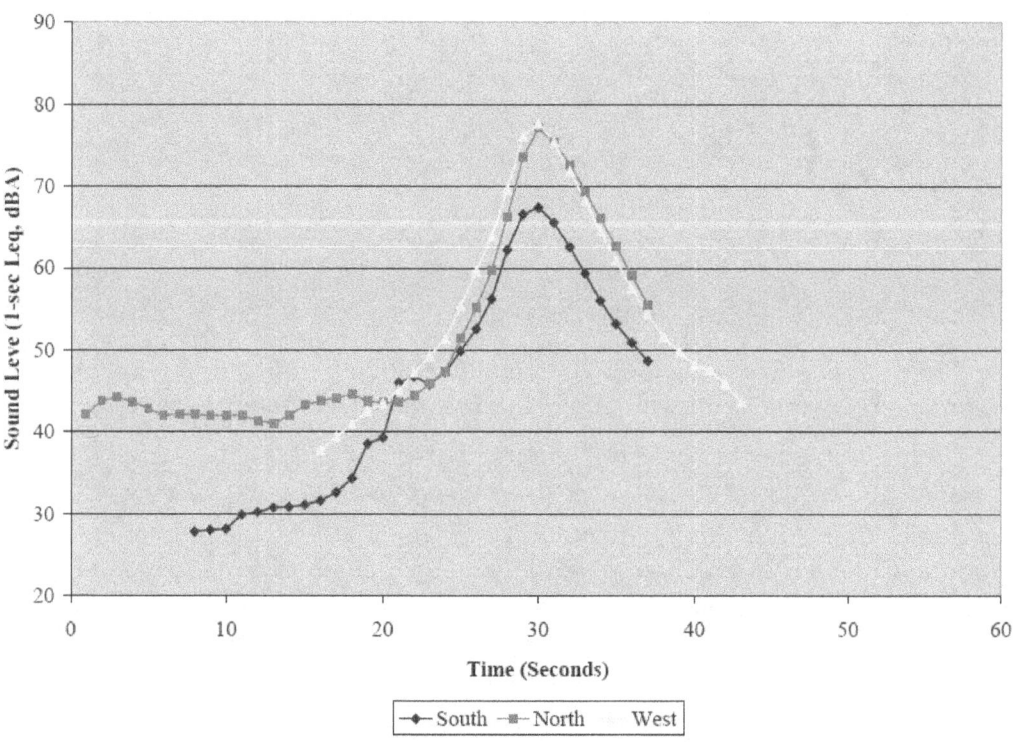

Figure 102. Yellowstone Snowcoach – SNOVAN5, High Speed Time History Comparison

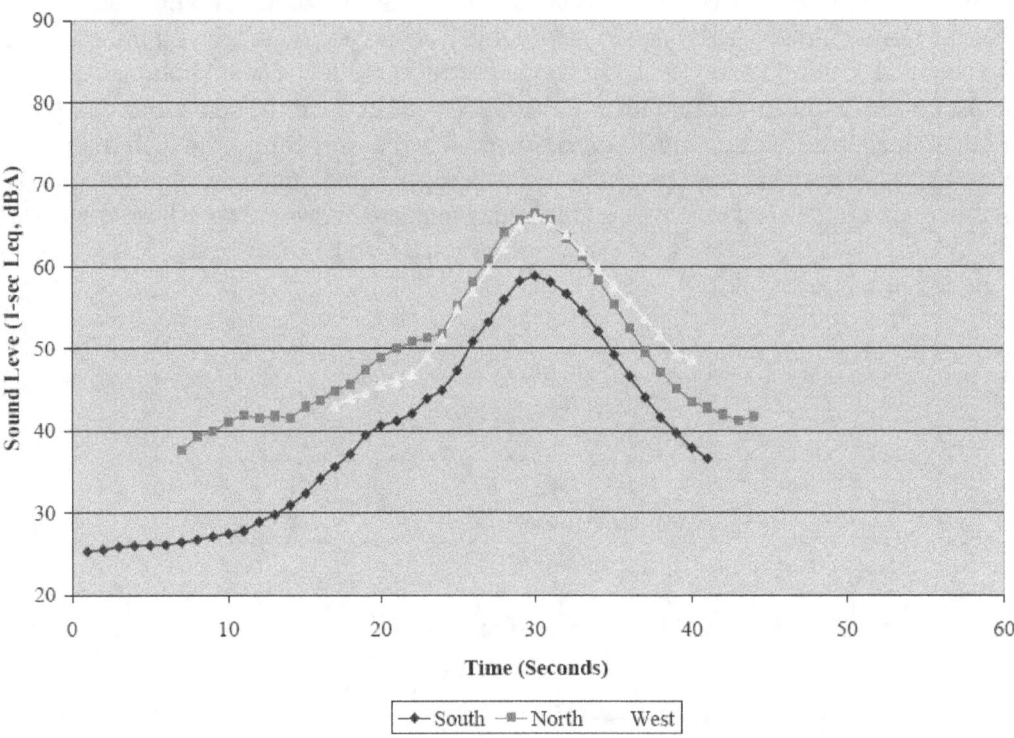

Figure 103. Yellowstone Snowcoach – SNOVAN5, Low Speed Time History Comparison

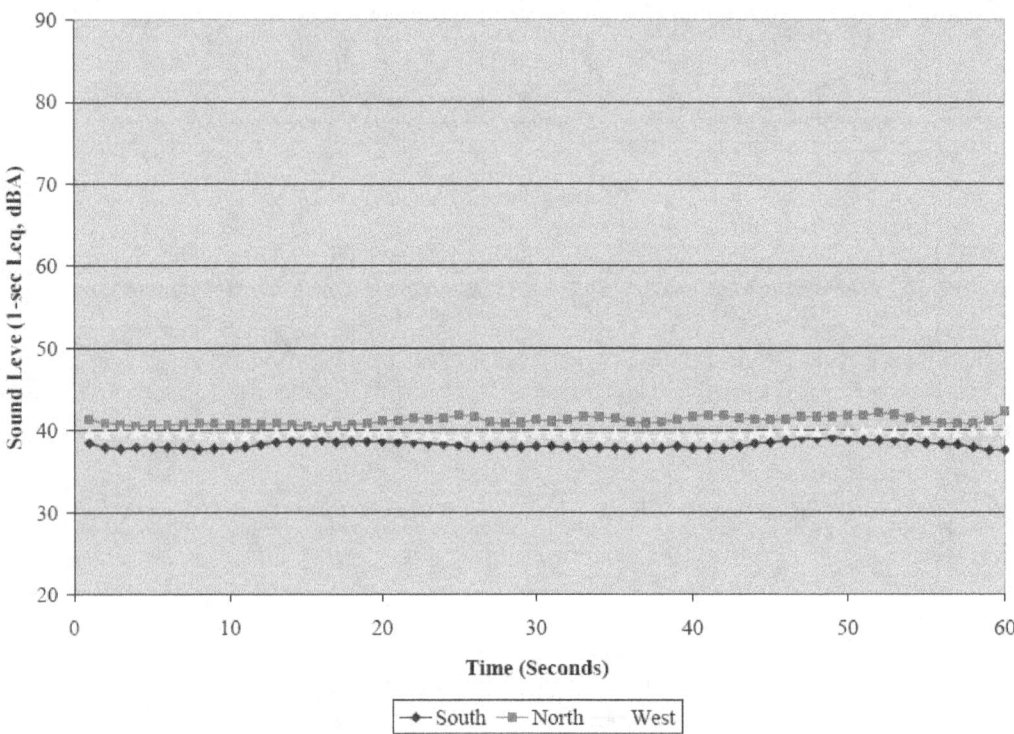

Figure 104. Yellowstone Snowcoach – SNOVAN5, Idle Time History Comparison

Examining Yellowstone Expedition – Hayden, Figure 105 though Figure 107, it should be noted that the north entrance had higher ambient sound levels which are apparent in the time history plots. The high speed events show that the west entrance was the loudest; this can be attributed to the fact that the south and north entrances had a speed of approximately 20 mph, while the west entrance was at a speed of approximately 30 mph. As in the previous example speed differences alone cannot account for all site-to-site differences as can be seen by the low speed data comparison. Where speed differences were small the snow berm should also be considered. The snow berms is discussed in section 4.3.2.

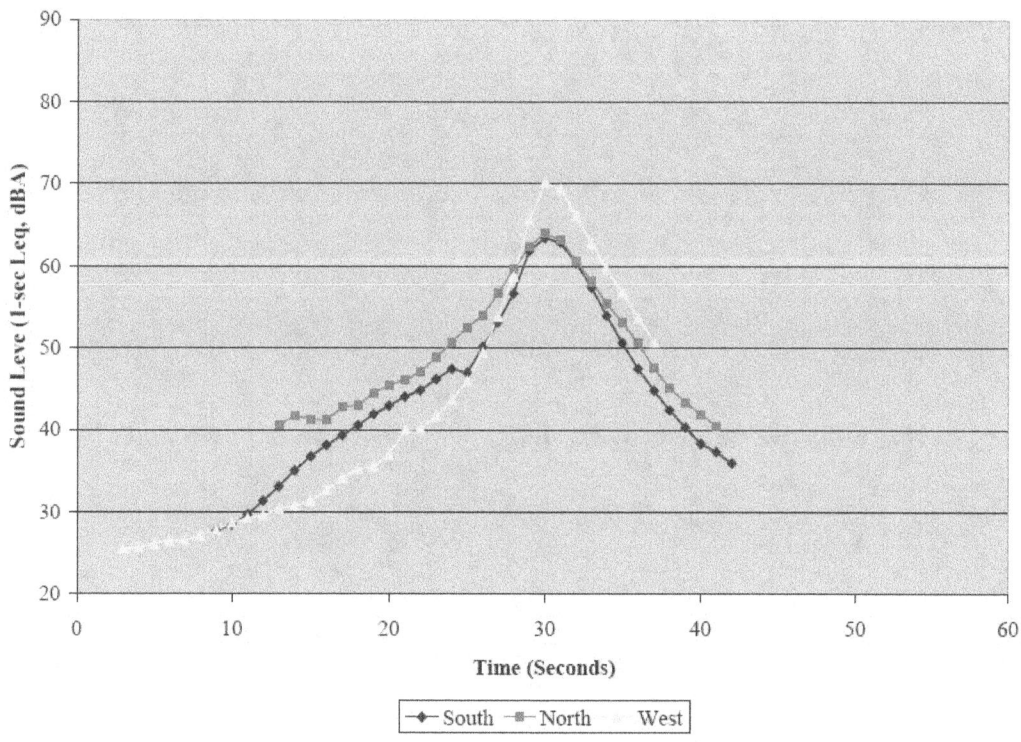

Figure 105. Yellowstone Expedition – Hayden, High Speed Time History Comparison

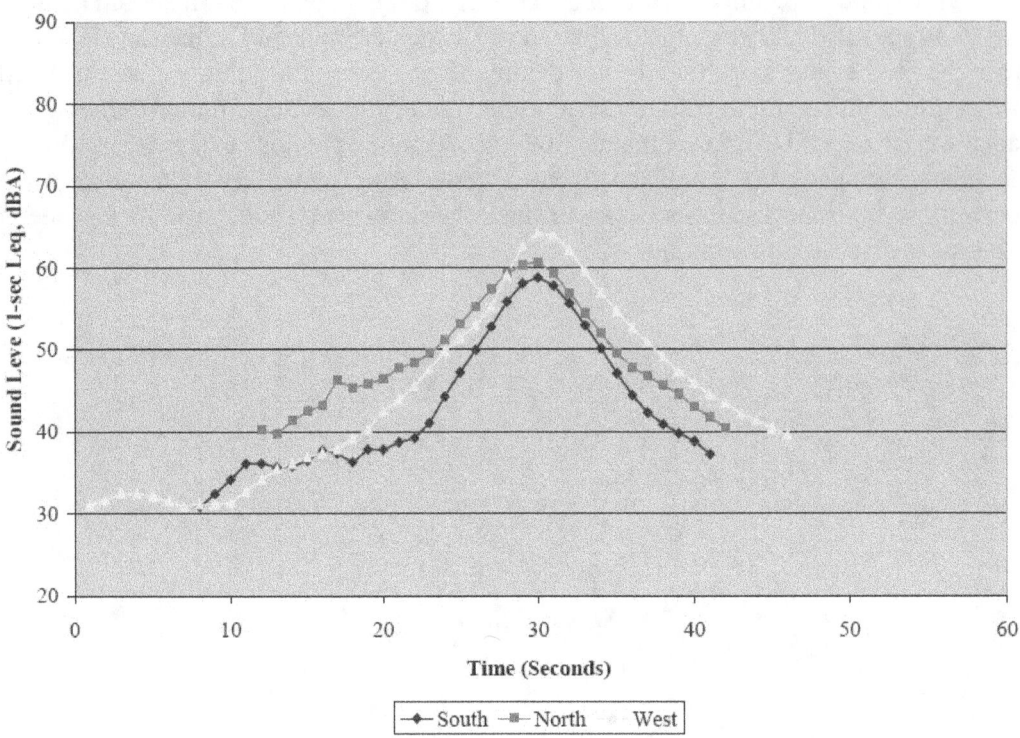

Figure 106. Yellowstone Expedition – Hayden, Low Speed Time History Comparison

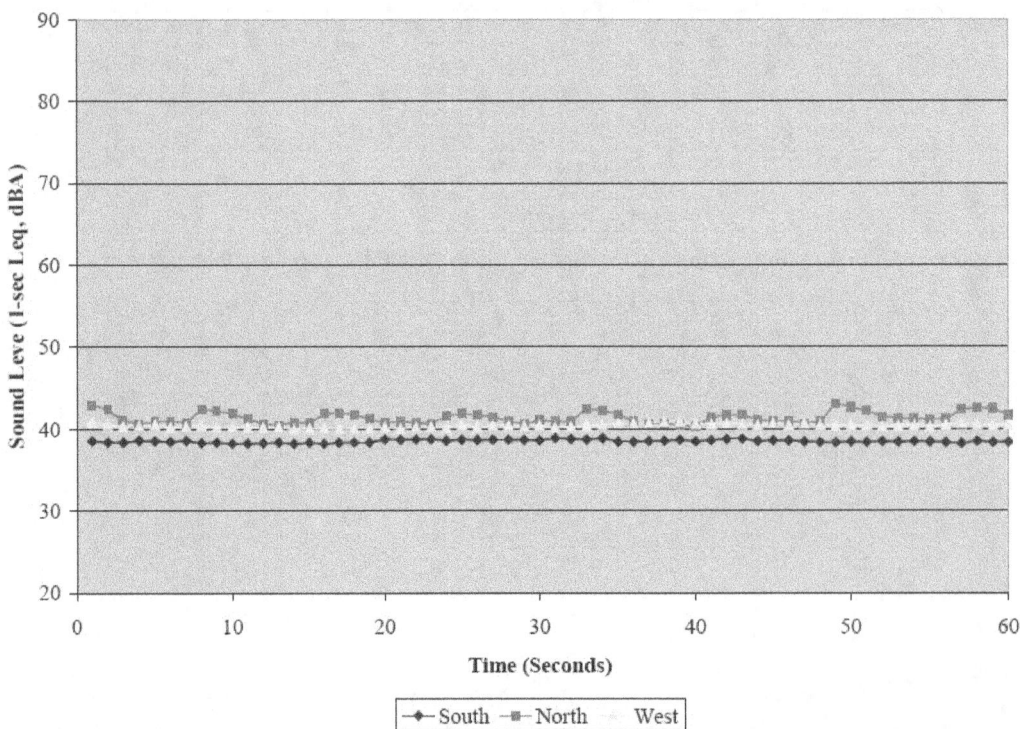

Figure 107. Yellowstone Expedition – Hayden, Idle Time History Comparison

Examining the Xanterra 713 in Figure 108 though Figure 110, it can be seen that both the high and low speed events line up closely between the three sites.

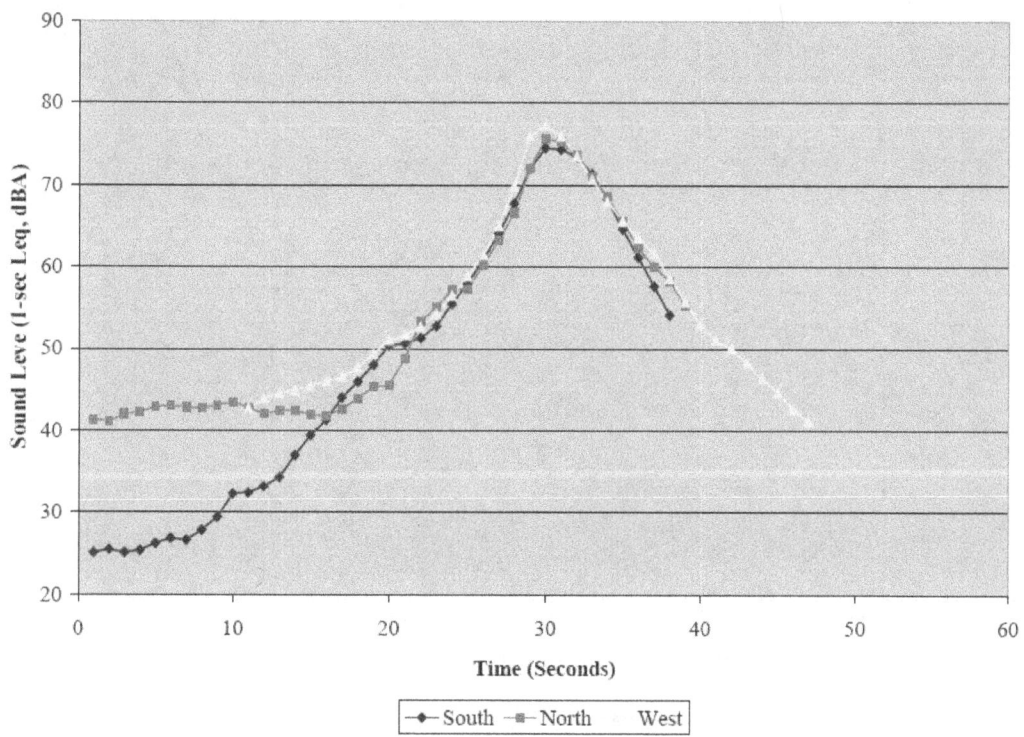

Figure 108. Xanterra 713, High Speed Time History Comparison

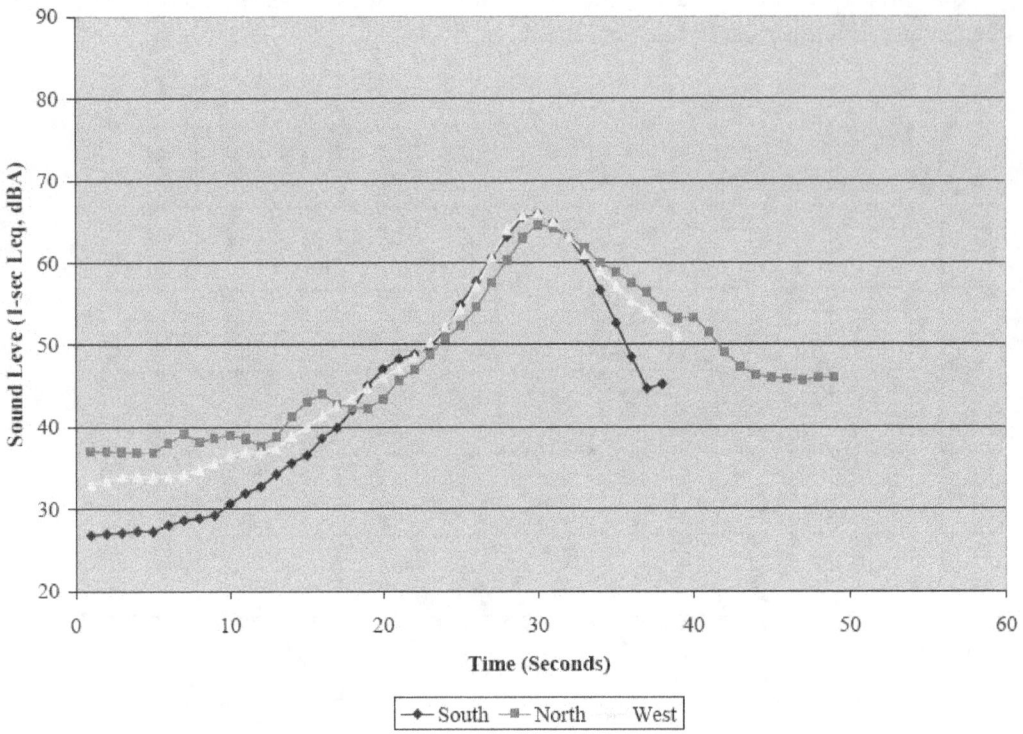

Figure 109. Xanterra 713, Low Speed Time History Comparison

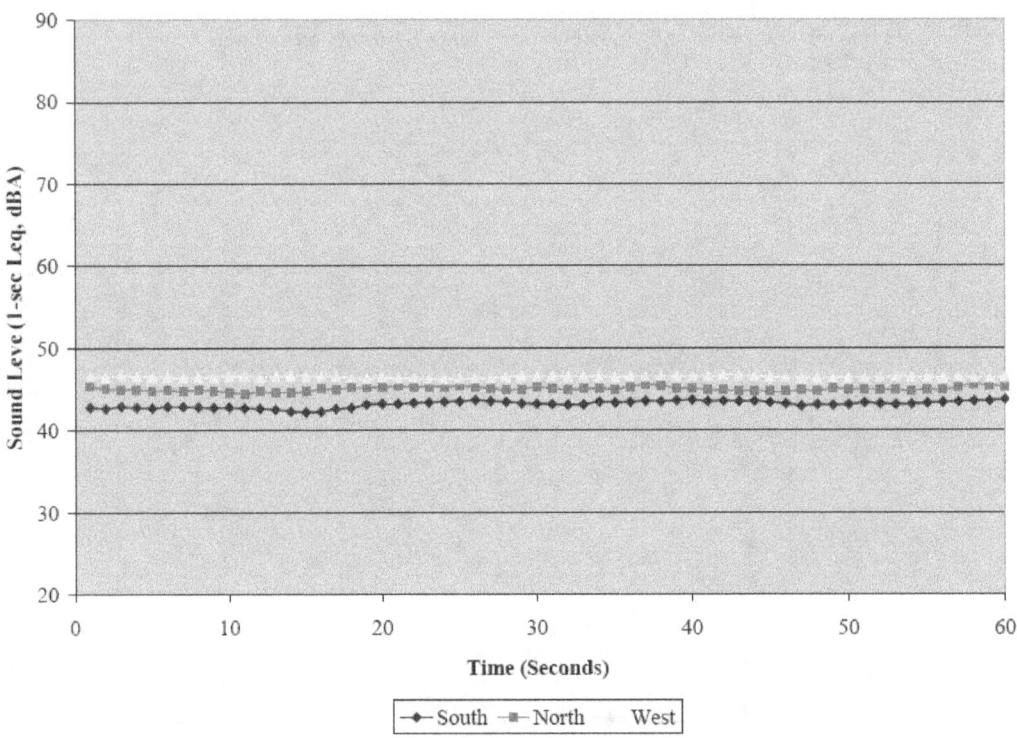

Figure 110. Xanterra 713, Idle Time History Comparison

Appendix H: L_{ASmx} Noise-Distance Curves

The L_{ASmx} noise-distance curve numerical values for all events can be seen in Table 56 through Table 61. Figure 111 though Figure 126 shows this data graphically.

Table 56. L_{ASmx} Noise-Distance Curve Values (part 1)

Vehicle	Xanterra 713			Yellowstone Expedition – Hayden		
Speed (mph)	15	25	32	15	20	30
Distance (Feet)	Maximum Sound Pressure Level (dBA)					
200	53.2	62.9	65.1	48.8	51.3	57.7
400	46.8	56.4	58.6	42.4	44.8	51.3
630	42.4	51.9	54.1	38.1	40.4	47.1
1000	37.8	47.2	49.3	33.6	35.8	42.6
2000	30.7	39.6	41.5	26.5	28.6	35.5
4000	23.0	31.3	32.7	19.1	21.1	27.7
6300	17.6	25.3	26.2	13.8	15.9	22.0
10000	11.8	18.6	18.8	8.1	10.3	15.7
16000	5.6	11.4	10.6	1.8	4.3	8.4
25000	-0.8	4.2	2.2	-4.6	-2.0	0.6

Table 57. L_{ASmx} Noise-Distance Curve Values (part 2)

Vehicle	Yellowstone Snowcoach – SNOVAN5			Xanterra 430		Xanterra 537	
Speed (mph)	15	25	35	15	23	10	16
Distance (Feet)	Maximum Sound Pressure Level (dBA)						
200	51.3	55.6	65.3	52.6	58.2	58.6	64.0
400	44.8	49.1	58.9	46.3	51.9	52.0	57.3
630	40.4	44.8	54.6	42.1	47.6	47.5	52.7
1000	35.8	40.2	50.0	37.7	43.1	42.8	47.7
2000	28.6	32.9	42.8	30.9	36.0	35.2	39.7
4000	20.7	25.1	34.9	23.4	28.3	27.3	31.0
6300	15.1	19.6	29.0	17.9	22.7	21.7	24.9
10000	8.9	13.5	22.3	11.7	16.4	15.6	18.2
16000	1.9	6.7	14.3	4.7	9.1	8.8	10.7
25000	-5.4	-0.5	5.3	-2.7	1.2	1.6	2.7

Table 58. L_{ASmx} Noise-Distance Curve Values (part 3)

Vehicle	Xanterra 707		Xanterra 709		Xanterra 710	
Speed (mph)	15	29	15	28	15	32
Distance (Feet)	Maximum Sound Pressure Level (dBA)					
200	55.3	67.7	55.3	67.1	51.9	63.8
400	48.8	61.2	48.1	60.1	45.4	57.3
630	44.3	56.8	43.1	55.3	40.9	52.8
1000	39.7	52.1	37.8	50.3	36.3	48.1
2000	32.2	44.5	29.6	42.2	29.0	40.5
4000	24.0	36.0	21.1	33.5	21.1	32.0
6300	18.3	29.6	15.2	27.4	15.5	25.8
10000	12.1	22.1	9.0	20.6	9.4	19.0
16000	5.5	13.5	2.5	13.0	2.6	11.5
25000	-1.1	4.7	-4.0	4.9	-4.5	3.6

Table 59. L_{ASmx} Noise-Distance Curve Values (part 4)

Vehicle	Alpen Guide – Kitty		Yellowstone Snowcoach – SNOVAN4		Yellowstone Expedition – Eleanor	
Speed (mph)	15	31	15	33	15	25
Distance (Feet)	Maximum Sound Pressure Level (dBA)					
200	47.0	54.2	50.6	63.3	52.4	56.2
400	40.6	47.7	44.2	56.9	46.0	49.8
630	36.2	43.3	39.9	52.6	41.6	45.4
1000	31.6	38.5	35.4	48.2	36.9	40.8
2000	24.5	31.0	28.5	41.2	29.3	33.5
4000	16.9	22.8	21.0	33.7	20.9	25.4
6300	11.7	17.0	15.8	28.4	14.9	19.5
10000	6.4	10.9	10.1	22.4	8.4	13.1
16000	0.8	4.3	3.7	15.6	1.3	5.9
25000	-4.8	-2.5	-2.9	8.1	-5.9	-1.6

Table 60. L_{ASmx} Noise-Distance Curve Values (part 5)

Vehicle	See Yellowstone Tours #6		See Yellowstone Tours #9		Buffalo Bus Touring #3	
Speed (mph)	15	35	15	36	15	27
Distance (Feet)	Maximum Sound Pressure Level (dBA)					
200	57.0	65.5	63.3	65.6	54.3	62.3
400	50.4	59.2	56.9	59.3	47.9	56.0
630	45.9	54.9	52.6	55.1	43.5	51.8
1000	41.2	50.5	48.0	50.7	39.0	47.4
2000	33.8	43.4	40.5	43.6	31.8	40.4
4000	25.8	35.6	31.9	35.8	24.0	32.7
6300	20.0	29.8	25.3	30.0	18.3	27.1
10000	13.7	23.2	17.6	23.2	11.8	20.6
16000	6.5	15.2	8.8	15.2	4.2	13.1
25000	-0.8	6.2	-0.4	6.3	-4.1	4.6

Table 61. L_{ASmx} Noise-Distance Curve Values (part 6)

Vehicle	Buffalo Bus Touring #4		Buffalo Bus Touring T2	
Speed (mph)	15	24	15	24
Distance (Feet)	Maximum Sound Pressure Level (dBA)			
200	52.6	58.9	54.3	60.1
400	45.9	52.2	47.9	53.8
630	41.3	47.6	43.6	49.6
1000	36.5	42.7	39.1	45.1
2000	28.8	34.8	32.0	38.2
4000	20.5	26.0	24.2	30.5
6300	14.7	19.7	18.5	24.9
10000	8.5	12.9	12.0	18.4
16000	1.7	5.4	4.5	11.0
25000	-5.3	-2.3	-3.5	2.7

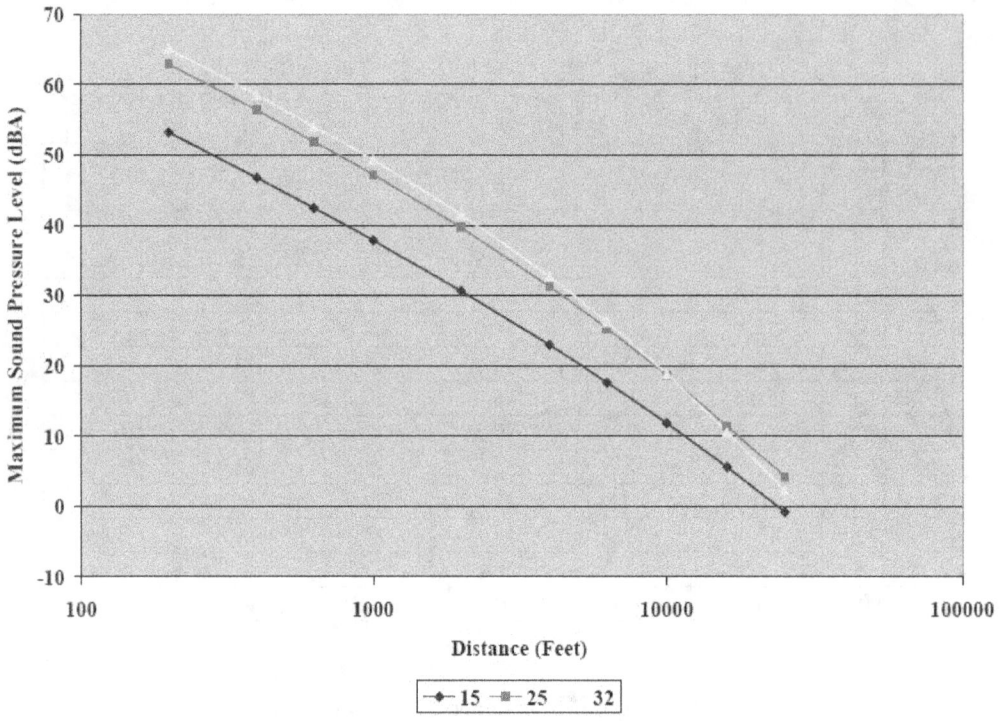

Figure 111. Xanterra L_{ASmx} Noise-Distance Curve

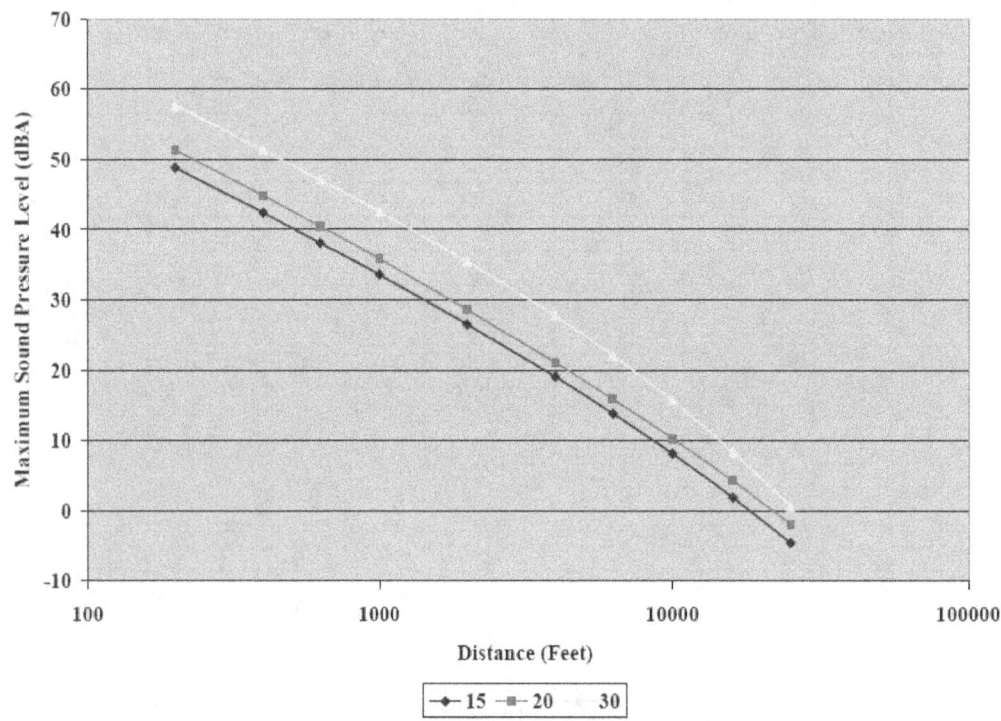

Figure 112. Yellowstone Expedition –Hayden L_{ASmx} Noise-Distance Curve

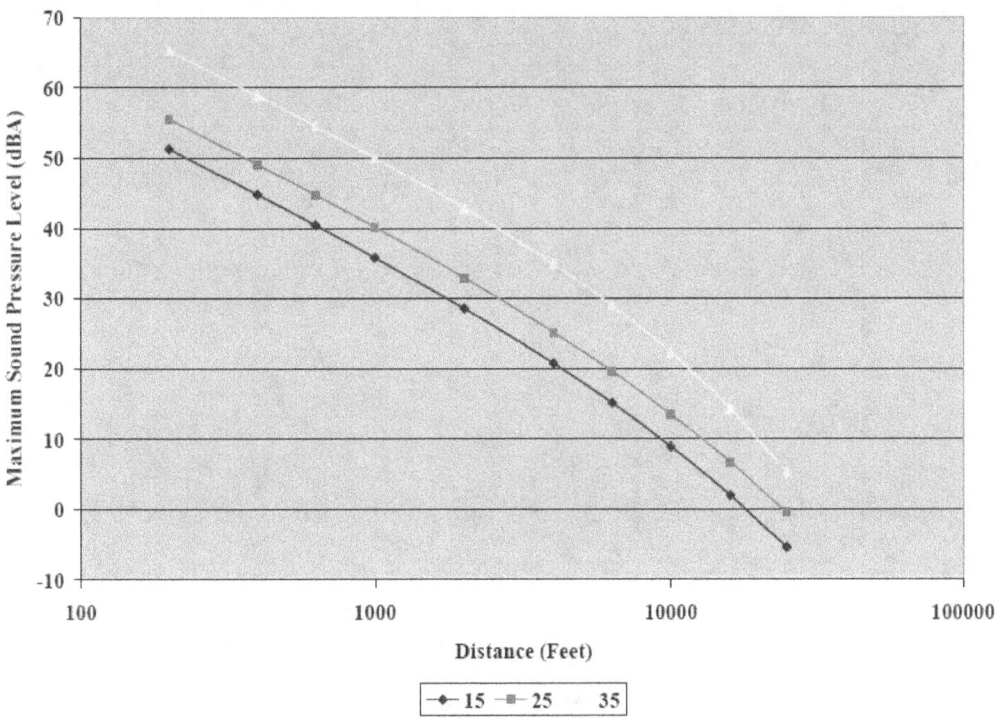

Figure 113. Yellowstone Snowcoach – SNOVAN5 L_{ASmx} Noise-Distance Curve

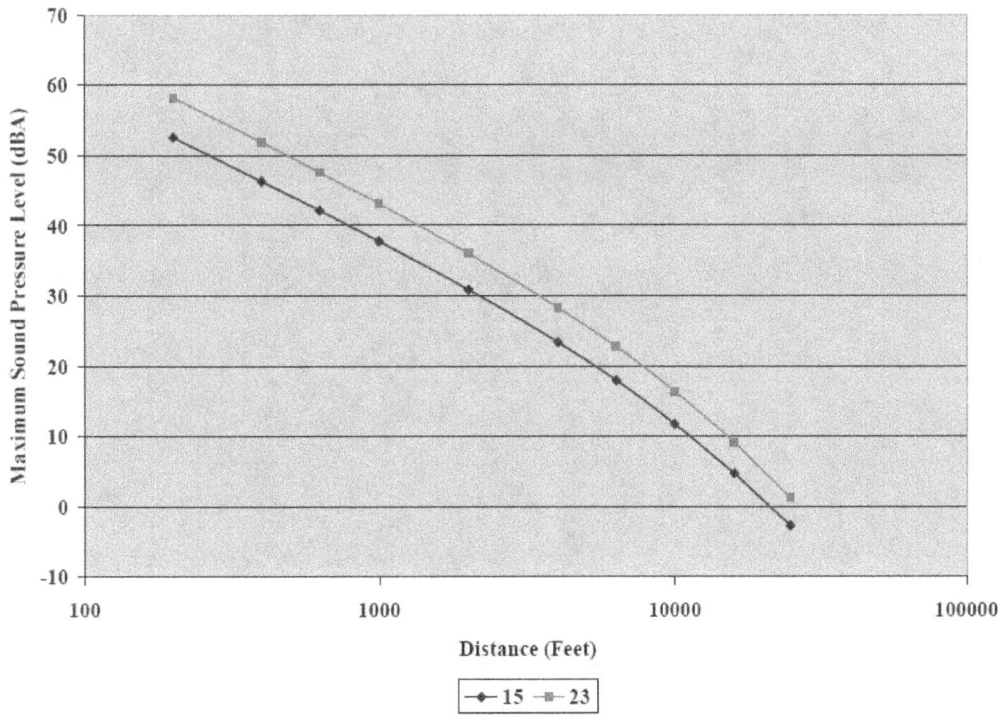

Figure 114. Xanterra 430 L_{ASmx} Noise-Distance Curve

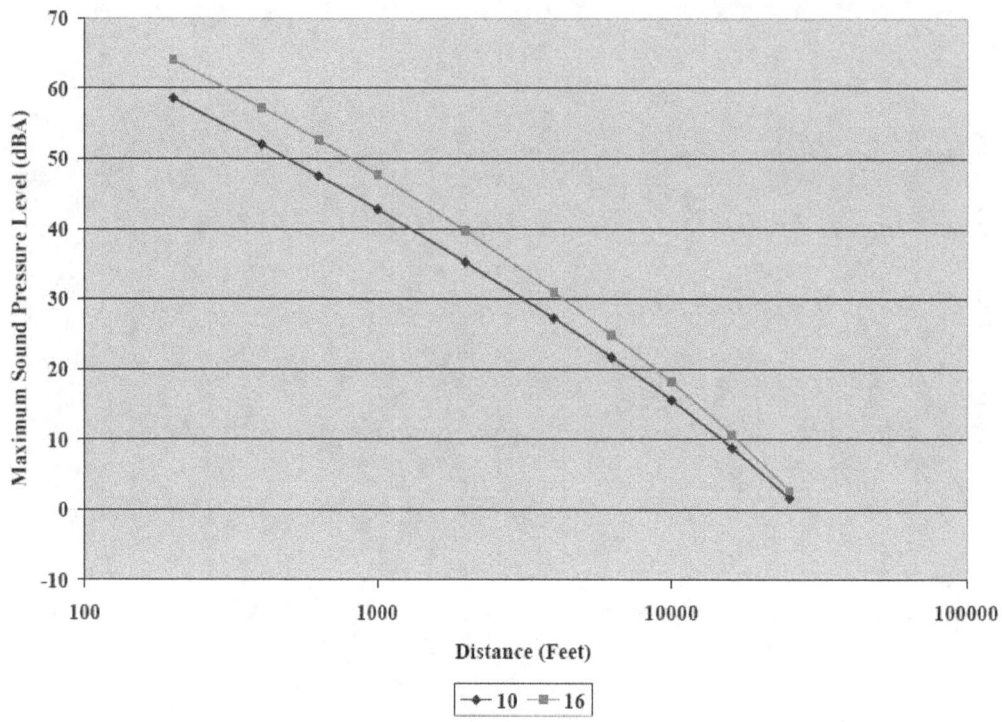

Figure 115. Xanterra 537 L$_{ASmx}$ Noise-Distance Curve

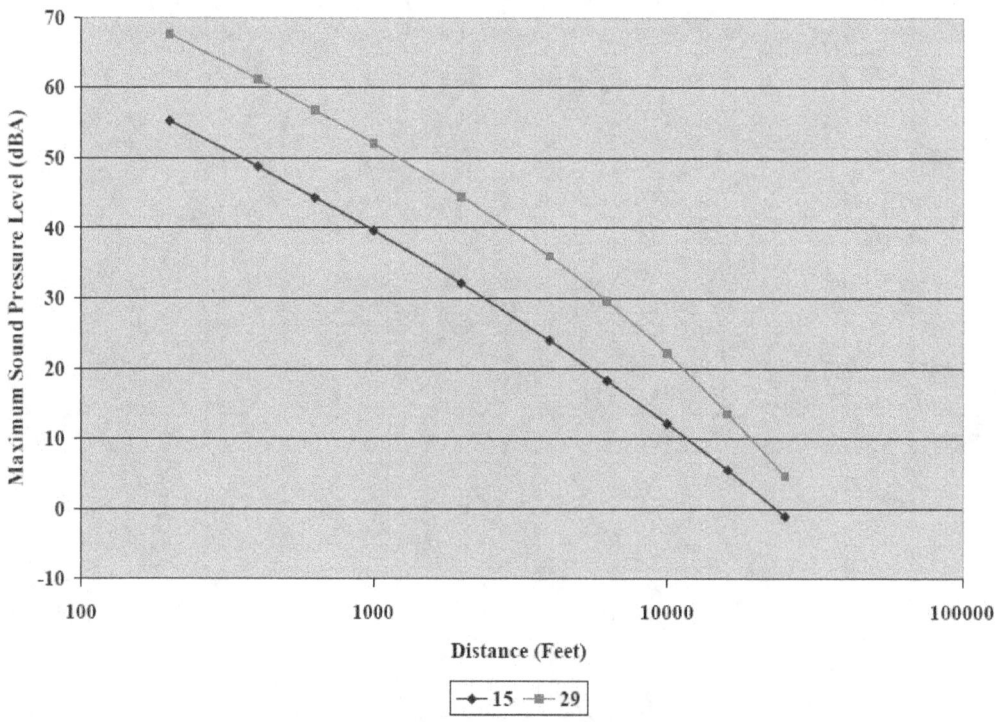

Figure 116. Xanterra 707 L$_{ASmx}$ Noise-Distance Curve

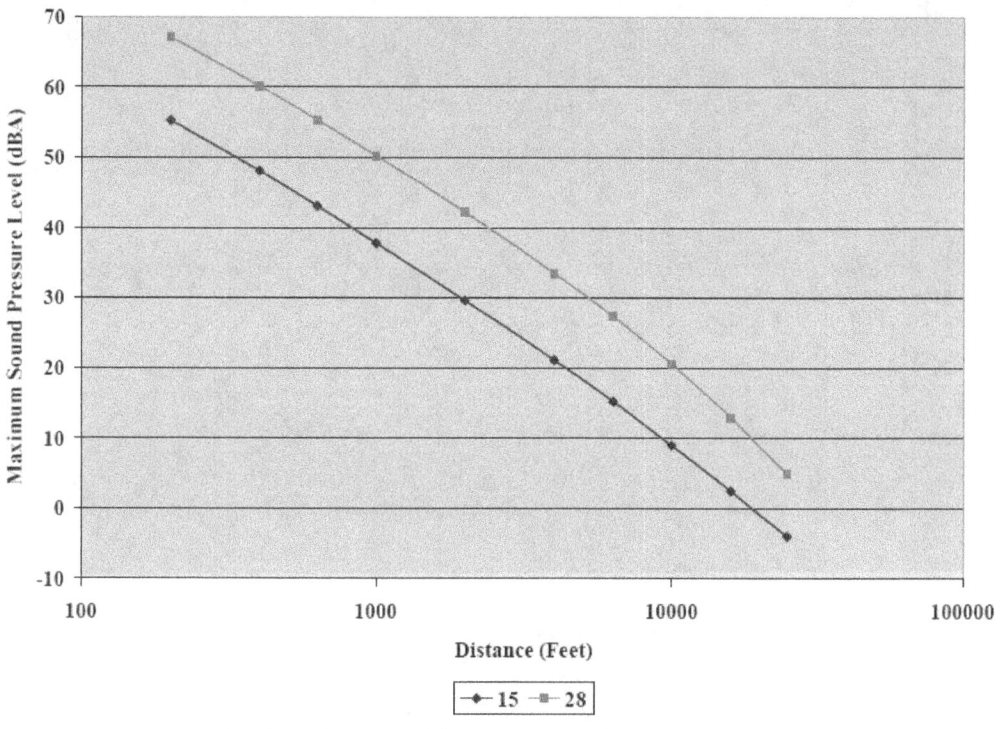

Figure 117. Xanterra 707 L_{ASmx} Noise-Distance Curve

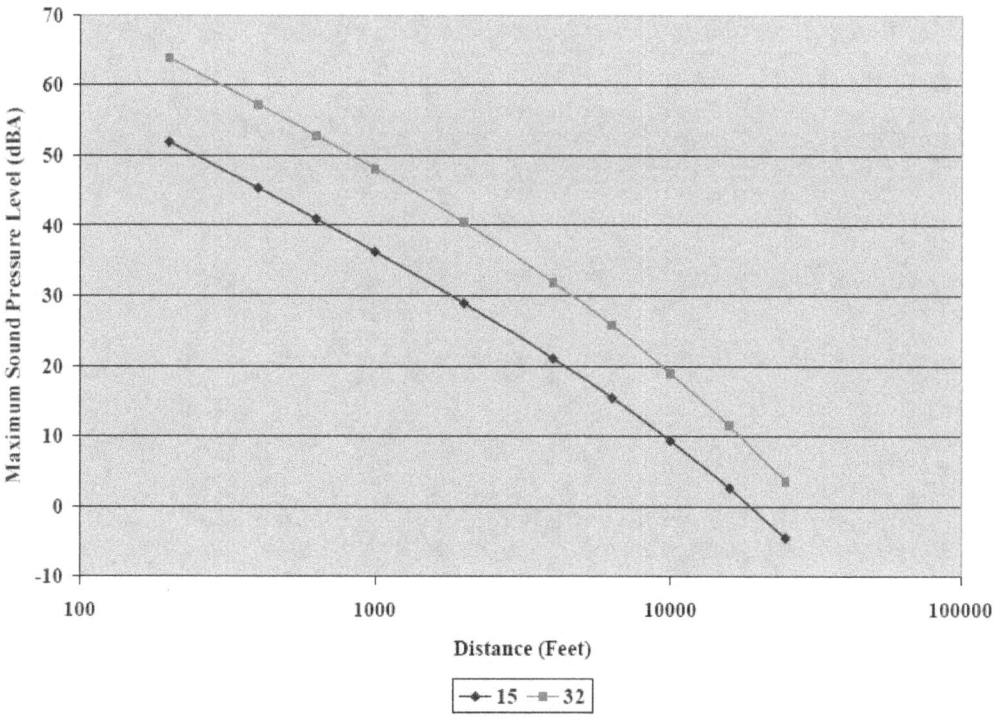

Figure 118. Xanterra 710 L_{ASmx} Noise-Distance Curve

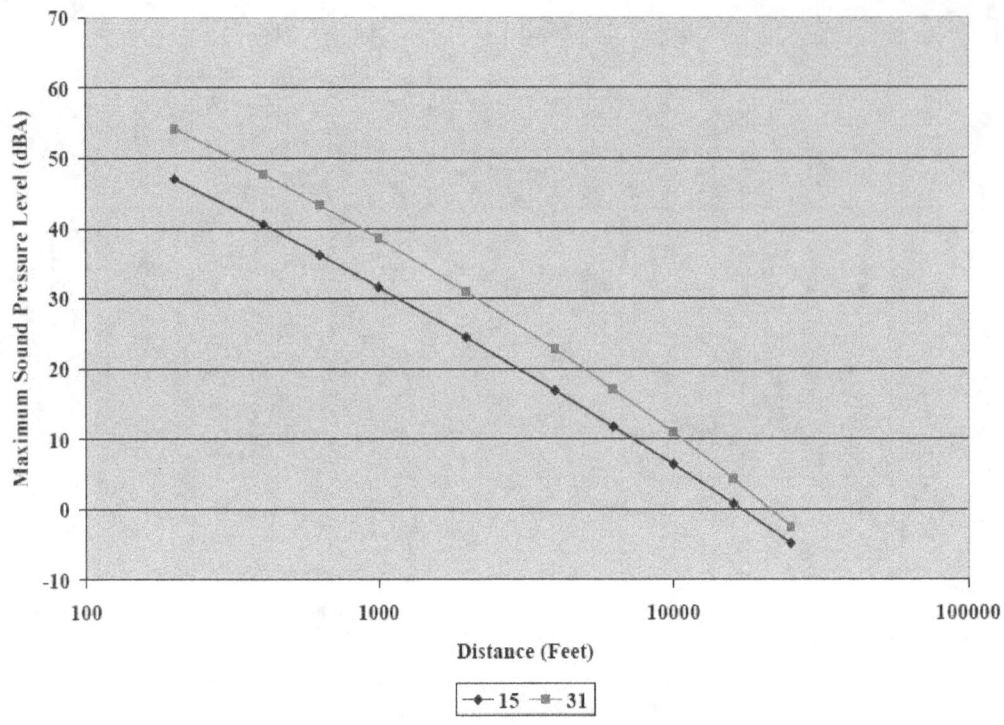

Figure 119. Alpen Guide – Kitty L_{ASmx} Noise-Distance Curve

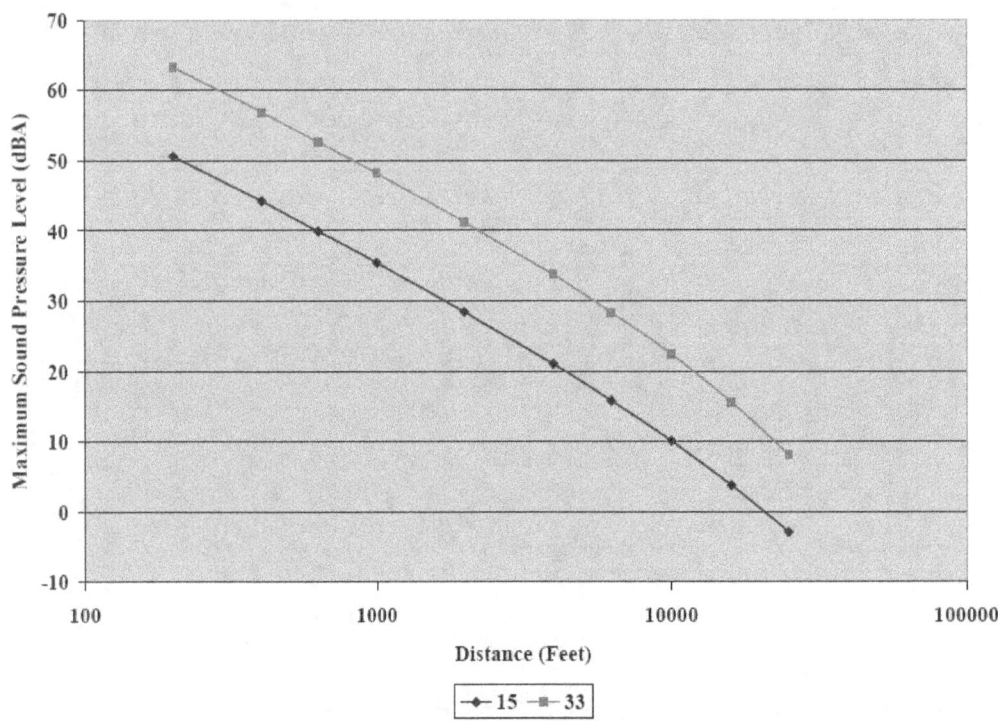

Figure 120. Yellowstone Snowcoach – SNOVAN4 L_{ASmx} Noise-Distance Curve

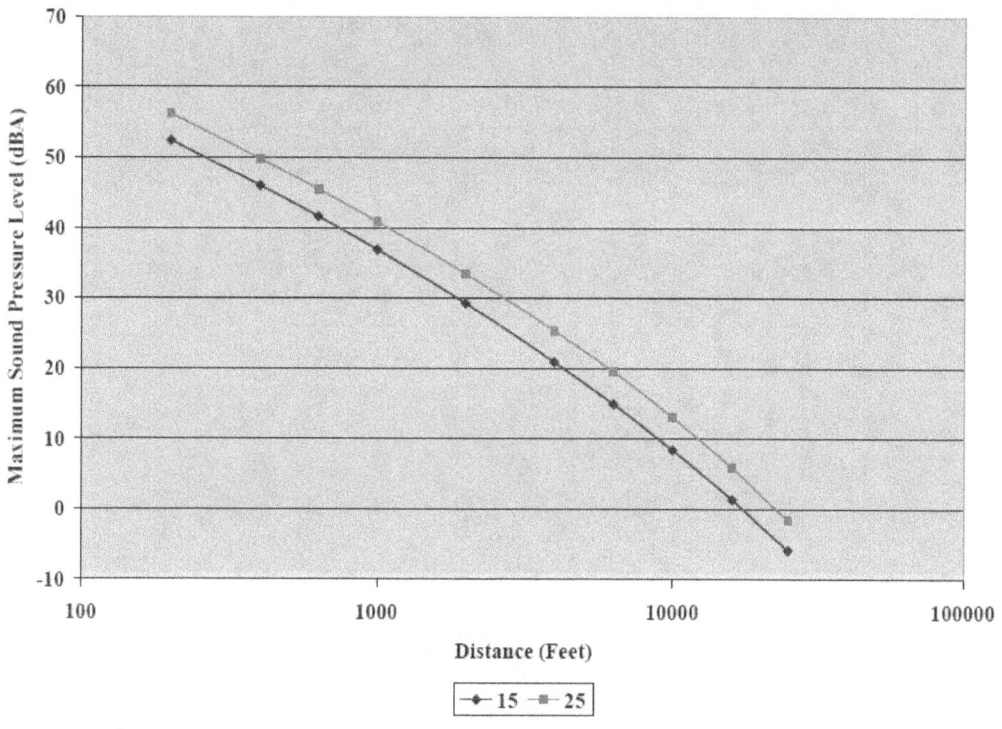

Figure 121. Yellowstone Expedition – Eleanor L_{ASmx} Noise-Distance Curve

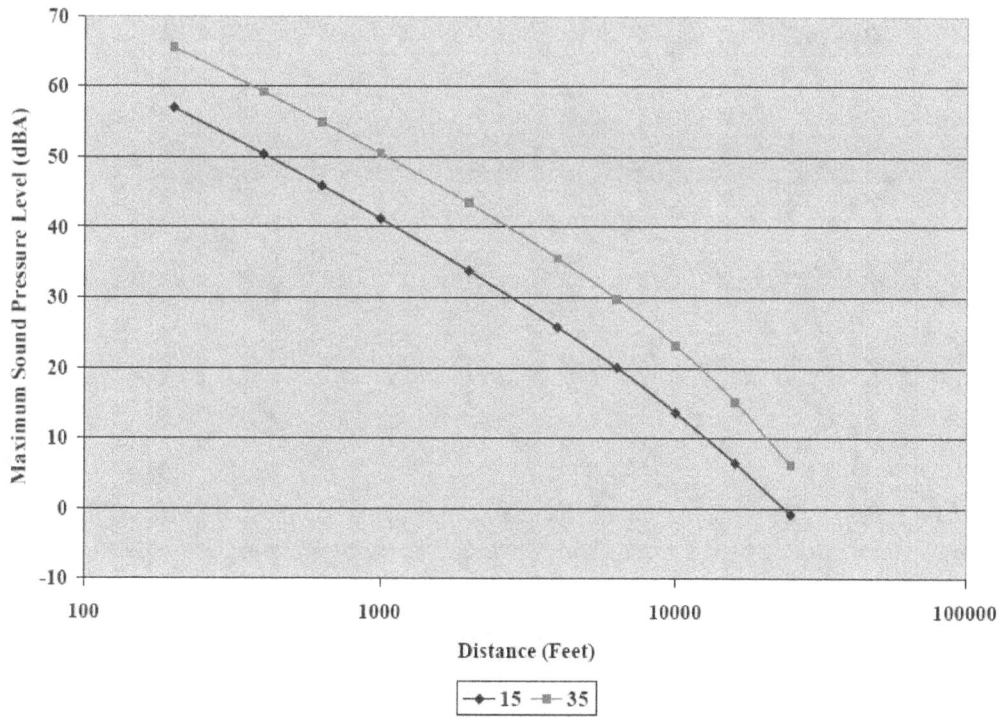

Figure 122. See Yellowstone Tours #6 L_{ASmx} Noise-Distance Curve

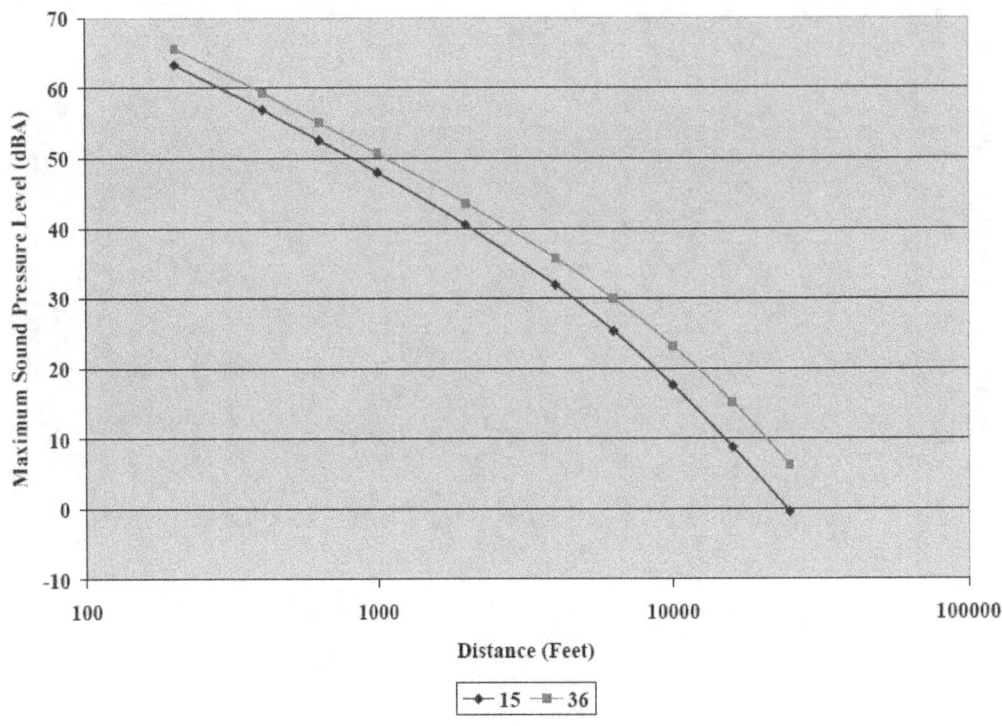

Figure 123. See Yellowstone Tours #9 L_{ASmx} Noise-Distance Curve

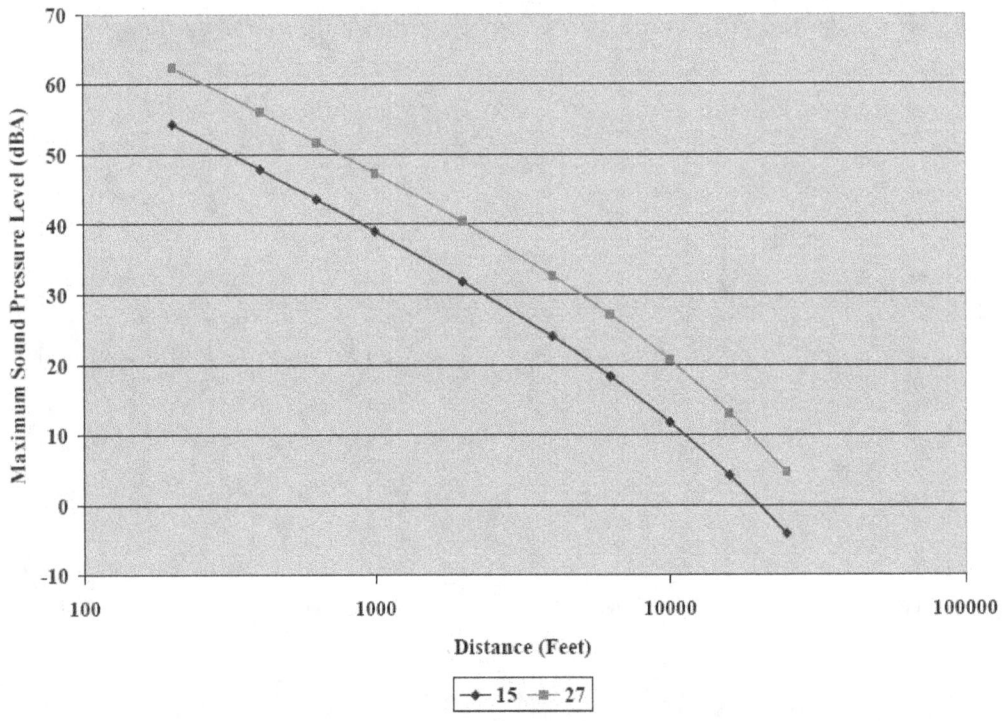

Figure 124. Buffalo Bus Touring #3 L_{ASmx} Noise-Distance Curves

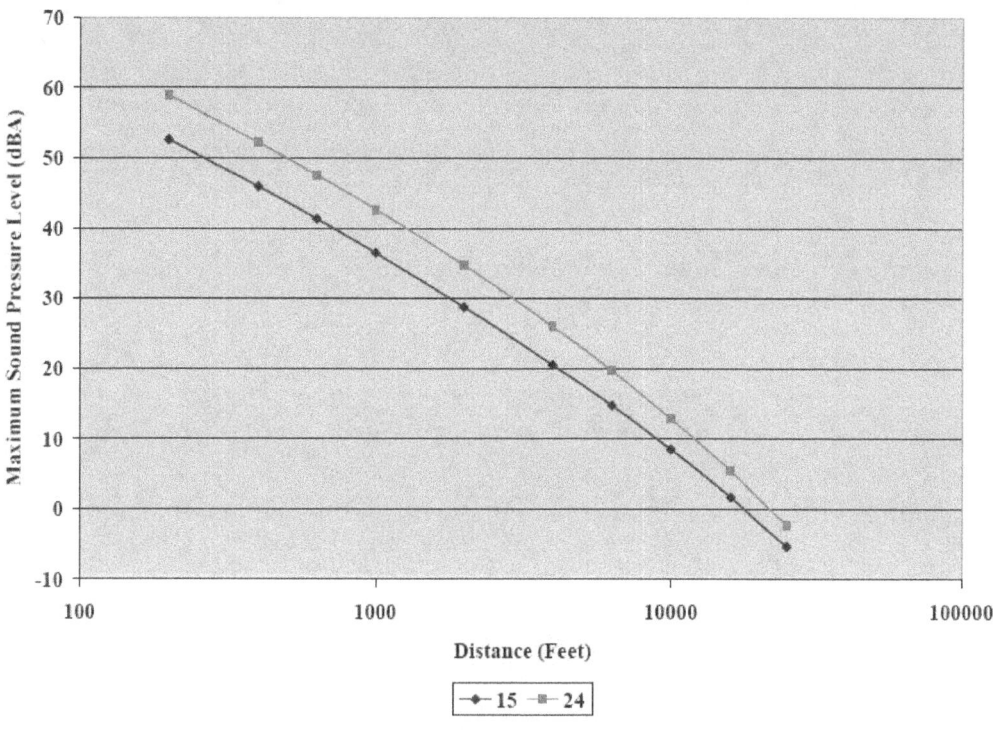

Figure 125. Buffalo Bus Touring #4 L_{ASmx} Noise-Distance Curves

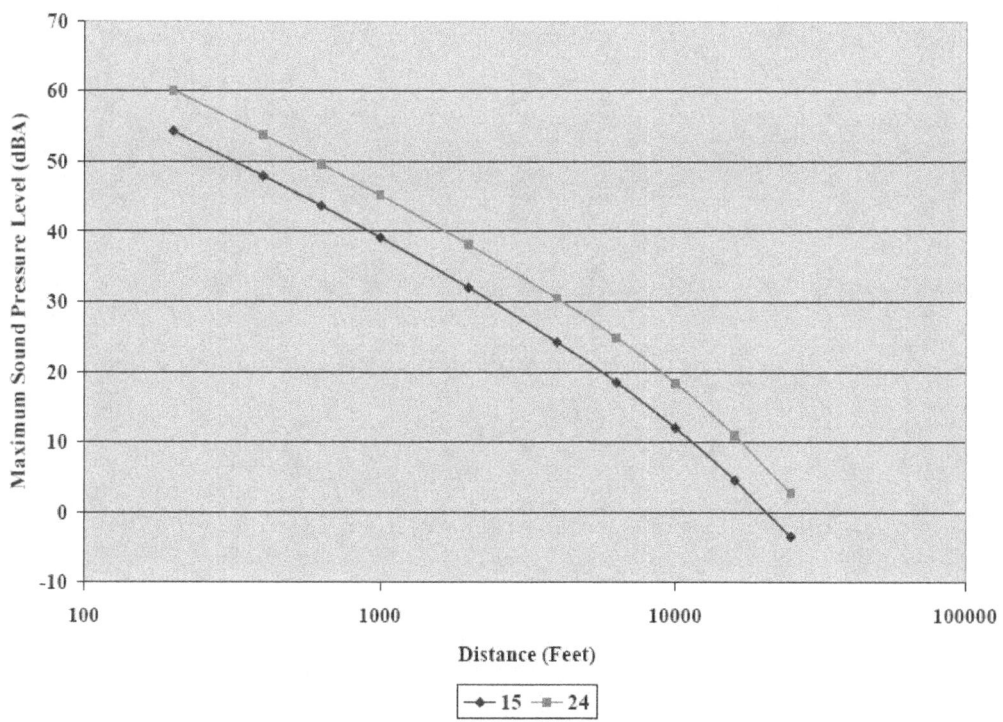

Figure 126. Buffalo Bus Touring T2 L_{ASmx} Noise-Distance Curve

Appendix I: Low Speed and High Speed SEL Noise-Distance Curves

The SEL noise-distance curve numerical values for all events for can be seen in Table 62 and Table 67. Figure 127 through Figure 142 show these data graphically.

Table 62. SEL Noise-Distance Curve Values (part 1)

Vehicle	Xanterra 713			Yellowstone Expedition – Hayden		
Speed (mph)	15	25	32	15	20	30
Distance (Feet)	Sound Exposure Level (dB)					
200	54.3	64.8	67.5	49.3	52.3	59.0
400	50.1	60.5	63.2	45.2	48.0	54.9
630	47.2	57.5	60.2	42.3	45.1	52.2
1000	44.1	54.3	56.9	39.3	42.0	49.2
2000	39.2	49.0	51.4	34.6	37.1	44.3
4000	33.8	42.9	44.8	29.3	31.8	38.8
6300	29.9	38.4	39.8	25.6	28.0	34.6
10000	25.6	33.2	33.9	21.4	24.0	29.8
16000	20.9	27.6	27.3	16.6	19.5	24.0
25000	15.9	21.8	20.3	11.7	14.7	17.7

Table 63. SEL Noise-Distance Curve Values (part 2)

Vehicle	Yellowstone Snowcoach – SNOVAN5			Xanterra 430		Xanterra 537	
Speed (mph)	15	25	35	15	23	10	16
Distance (Feet)	Sound Exposure Level (dB)						
200	51.9	56.9	67.0	53.2	59.8	61.0	66.4
400	47.7	52.7	62.9	49.2	55.7	56.6	62.0
630	44.8	49.8	60.1	46.5	52.9	53.6	58.8
1000	41.7	46.7	57.0	43.6	49.9	50.3	55.4
2000	36.7	41.8	52.1	39.0	45.1	45.1	49.6
4000	31.1	36.2	46.4	33.7	39.6	39.3	43.2
6300	27.0	32.1	42.0	29.8	35.5	35.3	38.5
10000	22.3	27.6	36.8	25.1	30.7	30.7	33.4
16000	16.8	22.3	30.3	19.6	24.9	25.4	27.4
25000	10.9	16.6	22.8	13.6	18.5	19.7	20.9

Table 64. SEL Noise-Distance Curve Values (part 3)

Vehicle	Xanterra 707		Xanterra 709		Xanterra 710	
Speed (mph)	15	29	15	28	15	32
Distance (Feet)	Sound Exposure Level (dB)					
200	56.8	69.6	56.0	68.1	53.1	66.2
400	52.6	65.4	51.1	63.4	48.8	62.0
630	49.6	62.4	47.5	60.1	45.8	59.0
1000	46.4	59.2	43.7	56.5	42.7	55.8
2000	41.2	53.9	37.8	50.7	37.6	50.4
4000	35.3	47.6	31.6	44.3	32.0	44.1
6300	31.1	42.7	27.2	39.6	27.9	39.4
10000	26.4	36.8	22.4	34.3	23.3	34.1
16000	21.4	29.7	17.4	28.2	18.0	28.2
25000	16.2	22.3	12.5	21.7	12.4	21.8

Table 65. SEL Noise-Distance Curve Values (part 4)

Vehicle	Alpen Guide – Kitty		Yellowstone Snowcoach – SNOVAN4		Yellowstone Expedition – Eleanor	
Speed (mph)	15	31	15	33	15	25
Distance (Feet)	Sound Exposure Level (dB)					
200	47.6	55.9	51.3	64.8	52.7	57.3
400	43.4	51.6	47.1	60.7	48.5	53.2
630	40.5	48.7	44.3	57.9	45.6	50.3
1000	37.5	45.5	41.4	55.0	42.4	47.2
2000	32.5	40.2	36.6	50.3	37.1	42.1
4000	27.2	34.2	31.5	45.1	31.0	36.3
6300	23.5	29.9	27.7	41.2	26.5	31.9
10000	19.7	25.3	23.5	36.7	21.5	27.0
16000	15.6	20.3	18.7	31.4	15.9	21.3
25000	11.5	14.9	13.5	25.4	10.2	15.2

Table 66. SEL Noise-Distance Curve Values (part 5)

Vehicle	See Yellowstone Tours #6		See Yellowstone Tours #9		Buffalo Bus Touring #3	
Speed (mph)	15	35	15	36	15	27
Distance (Feet)	Sound Exposure Level (dB)					
200	57.3	67.3	62.6	67.4	55.4	63.7
400	53.0	63.3	58.5	63.3	51.3	59.7
630	50.0	60.5	55.7	60.6	48.4	57.0
1000	46.8	57.5	52.6	57.6	45.4	54.1
2000	41.7	52.7	47.3	52.9	40.5	49.3
4000	35.9	47.2	41.0	47.3	34.9	43.9
6300	31.6	42.9	35.9	43.0	30.7	39.8
10000	26.7	37.7	29.7	37.7	25.7	34.8
16000	21.1	31.3	22.4	31.2	19.6	28.8
25000	15.3	23.7	14.7	23.8	12.8	21.8

Table 67. SEL Noise-Distance Curve Values (part 6)

Vehicle	Buffalo Bus Touring #4		Buffalo Bus Touring T2	
Speed (mph)	15	24	15	24
Distance (Feet)	Sound Exposure Level (dB)			
200	53.4	59.9	55.1	61.2
400	49.0	55.4	51.0	57.1
630	45.9	52.3	48.2	54.4
1000	42.5	48.8	45.2	51.5
2000	37.1	43.2	40.3	46.7
4000	31.1	36.7	34.8	41.3
6300	26.8	31.8	30.5	37.2
10000	22.1	26.5	25.6	32.3
16000	16.8	20.6	19.6	26.3
25000	11.2	14.4	13.1	19.5

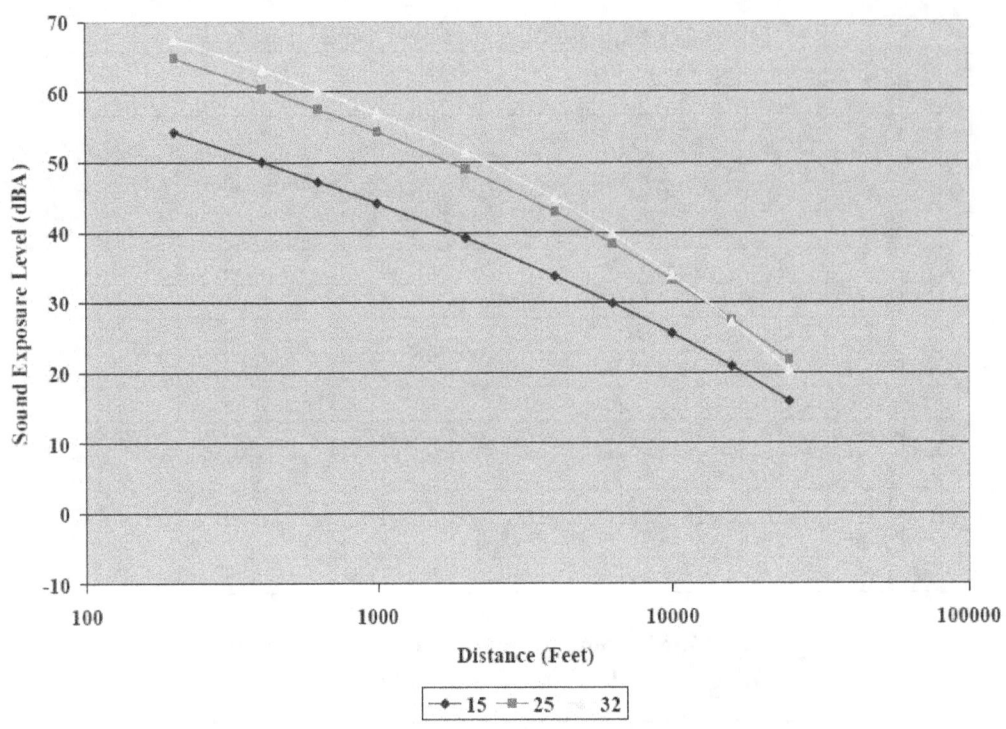

Figure 127. Xanterra 713 SEL Noise-Distance Curve

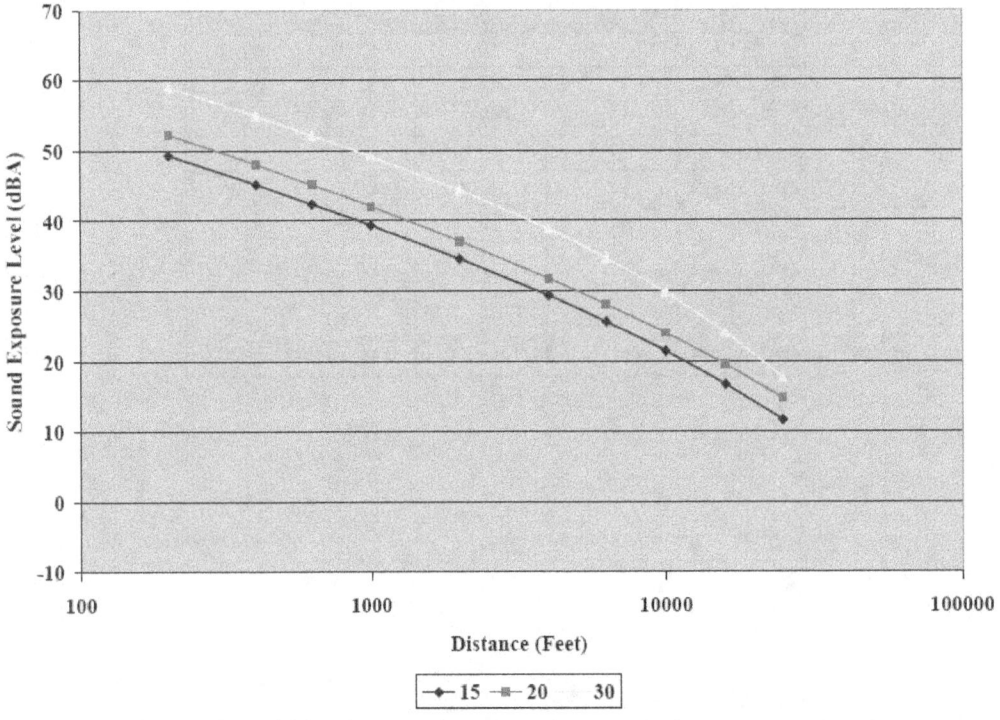

Figure 128. Yellowstone Expedition – Hayden SEL Noise-Distance Curve

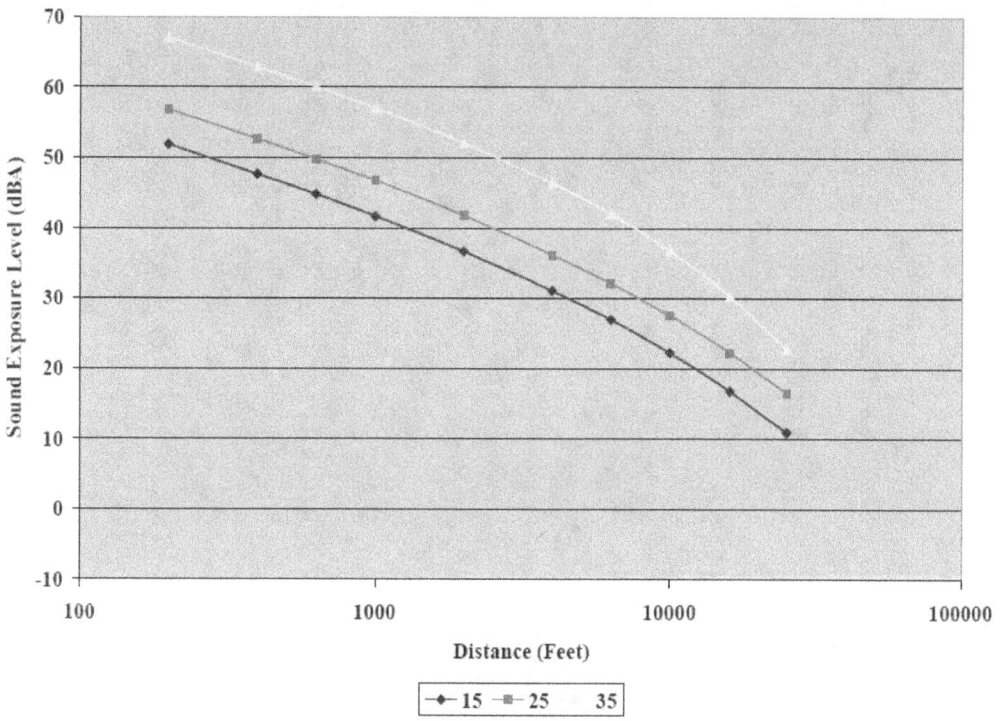

Figure 129. Yellowstone Snowcoach – SNOVAN5 SEL Noise-Distance Curve

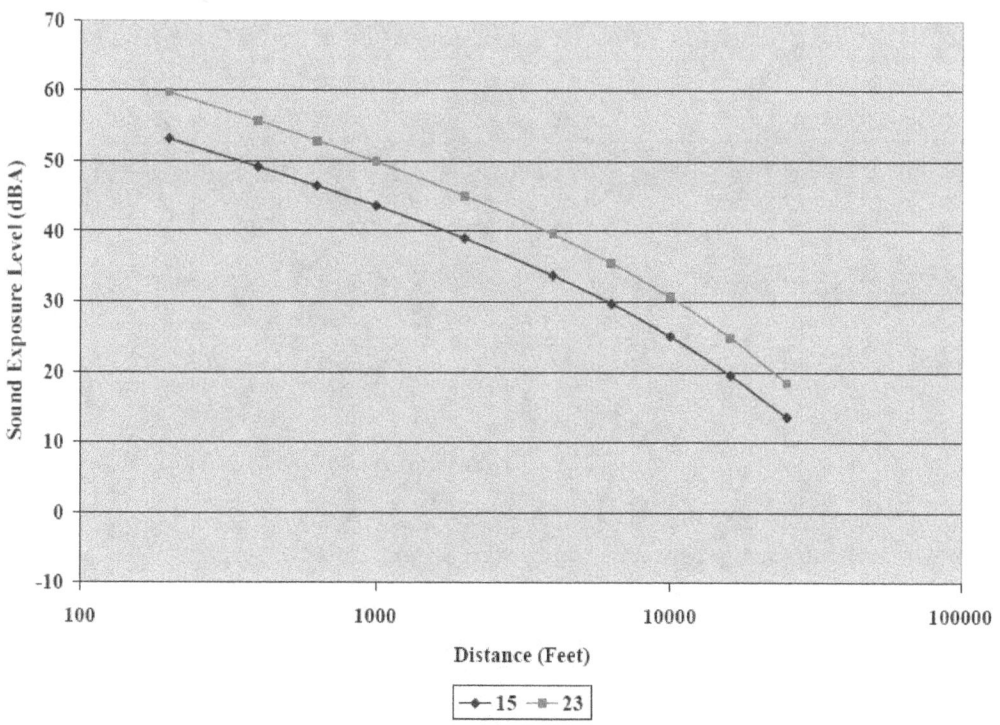

Figure 130. Xanterra 430 SEL Noise-Distance Curve

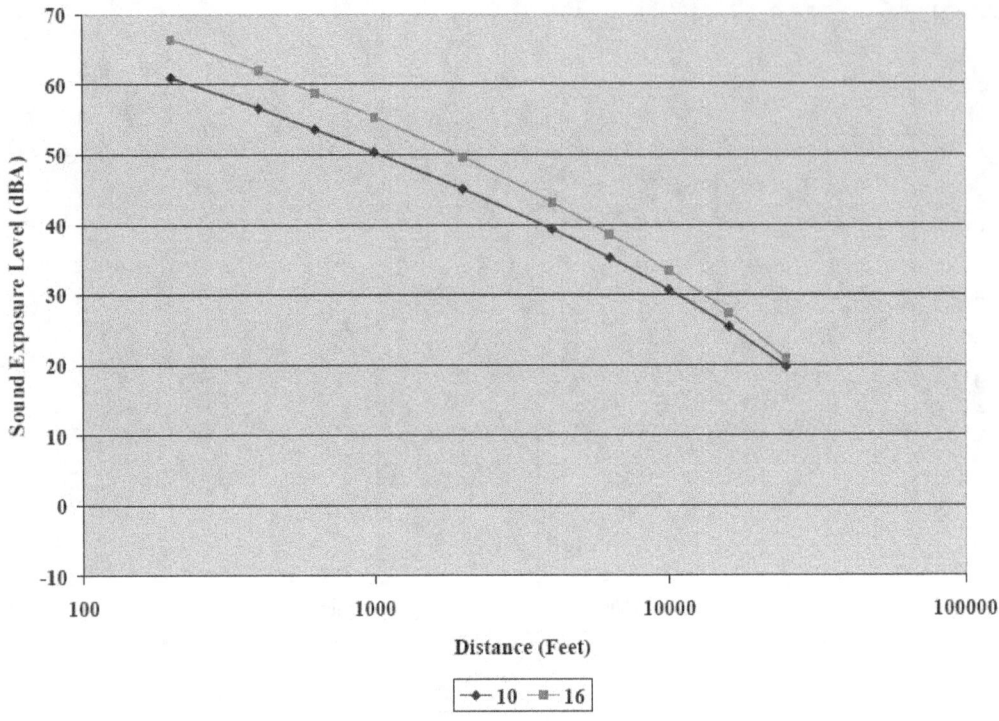

Figure 131. Xanterra 537 SEL Noise-Distance Curve

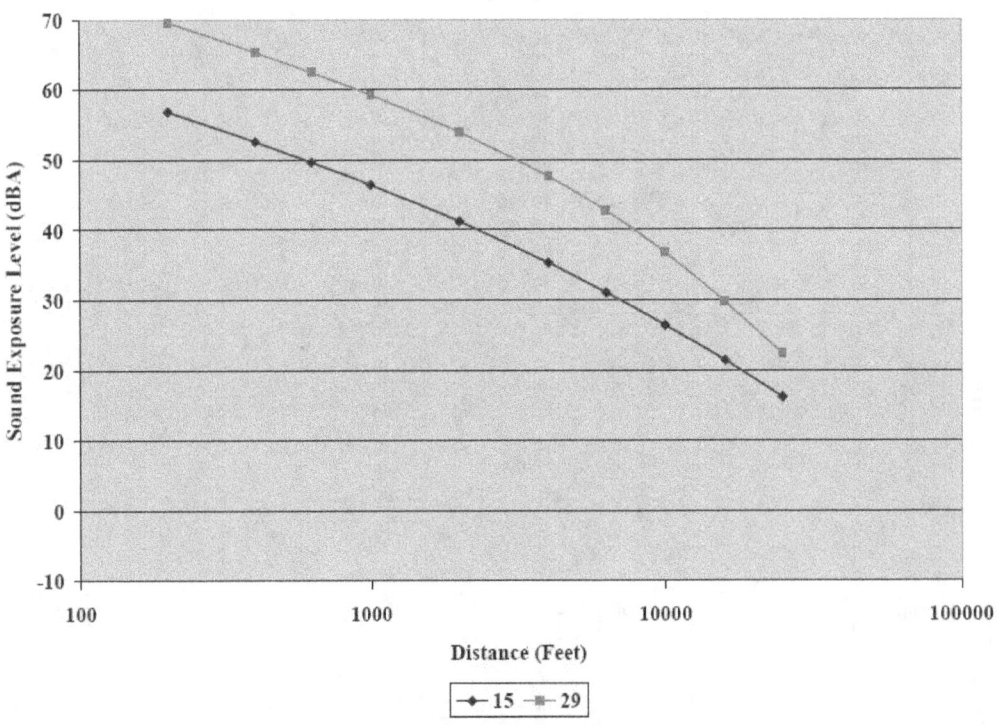

Figure 132. Xanterra 707 SEL Noise-Distance Curve

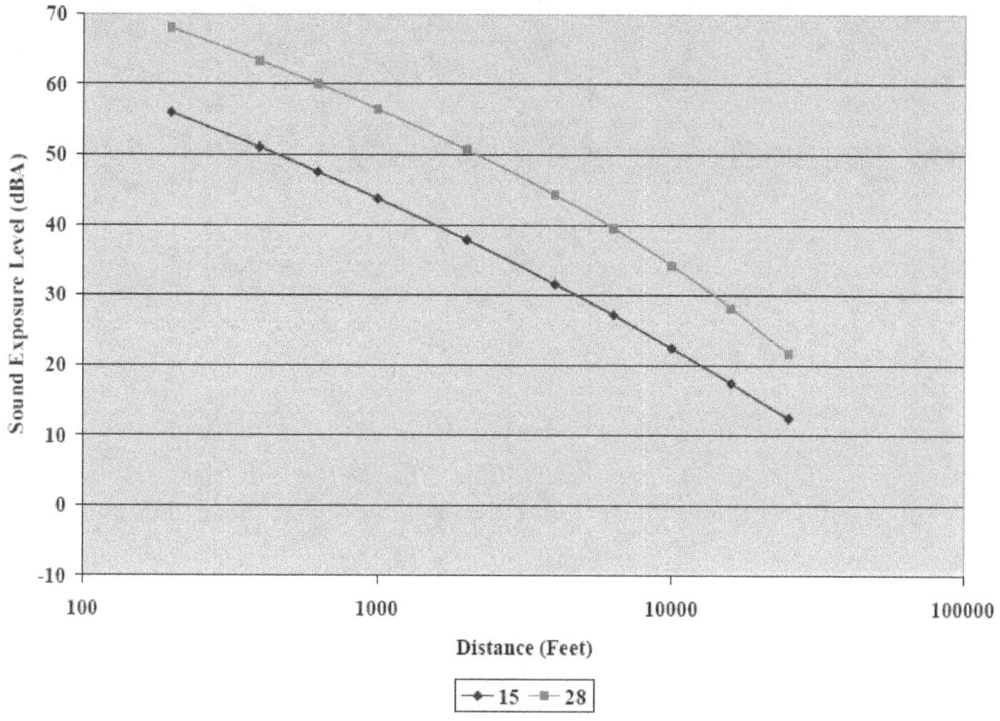

Figure 133. Xanterra 709 SEL Noise-Distance Curve

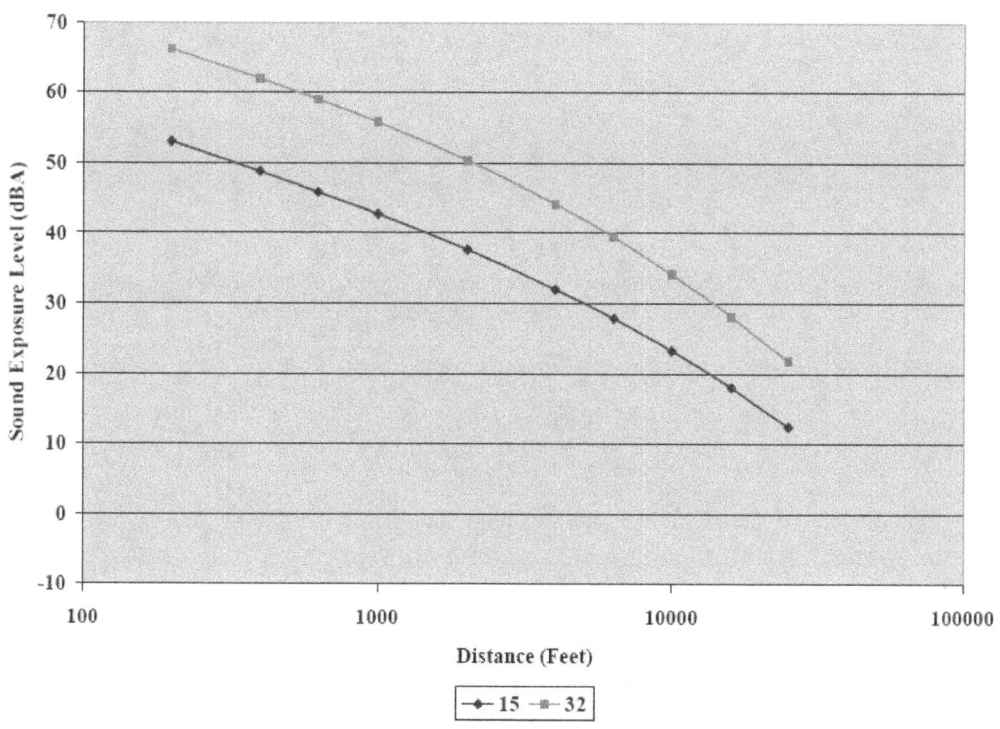

Figure 134. Xanterra 710 SEL Noise-Distance Curve

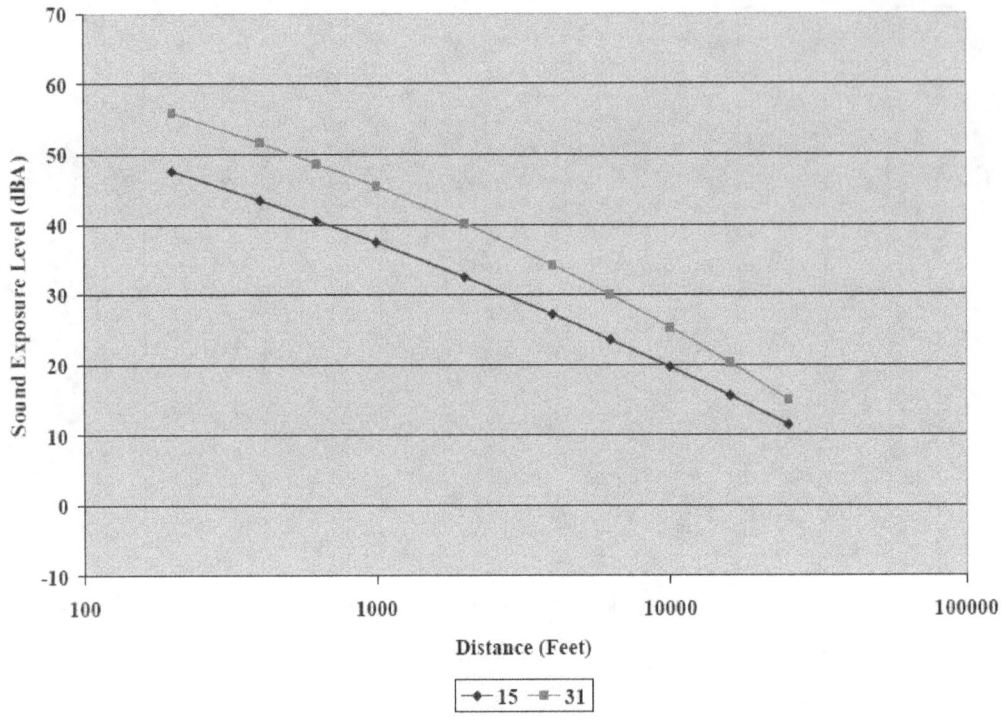

Figure 135. Alpen Guide – Kitty SEL Noise-Distance Curve

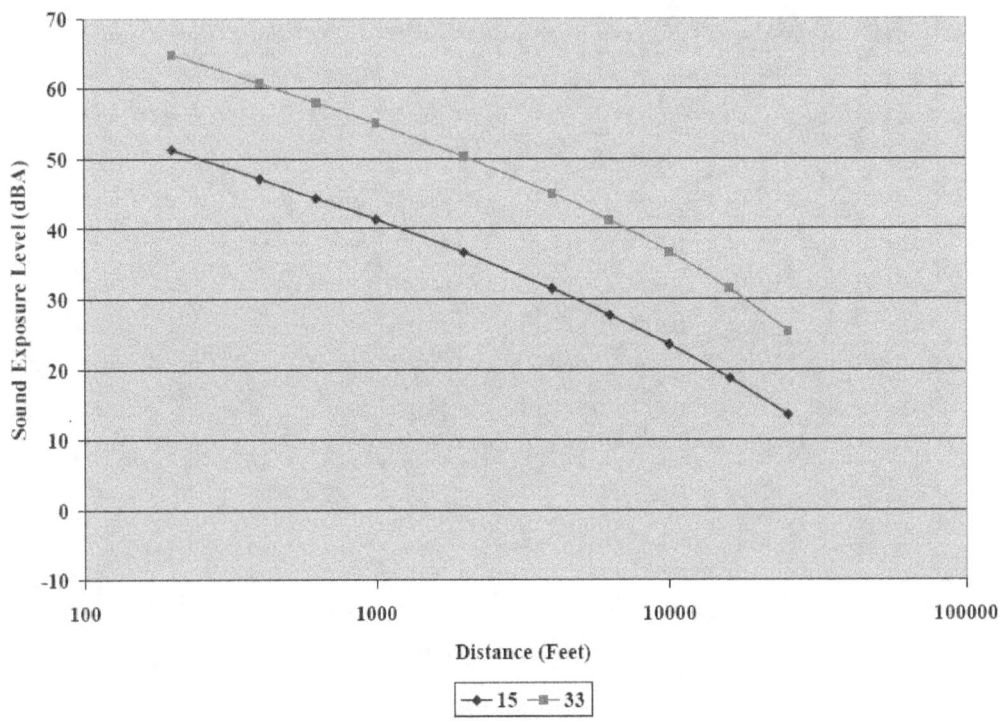

Figure 136. Yellowstone Snowcoach– SNOVAN4 SEL Noise-Distance Curve

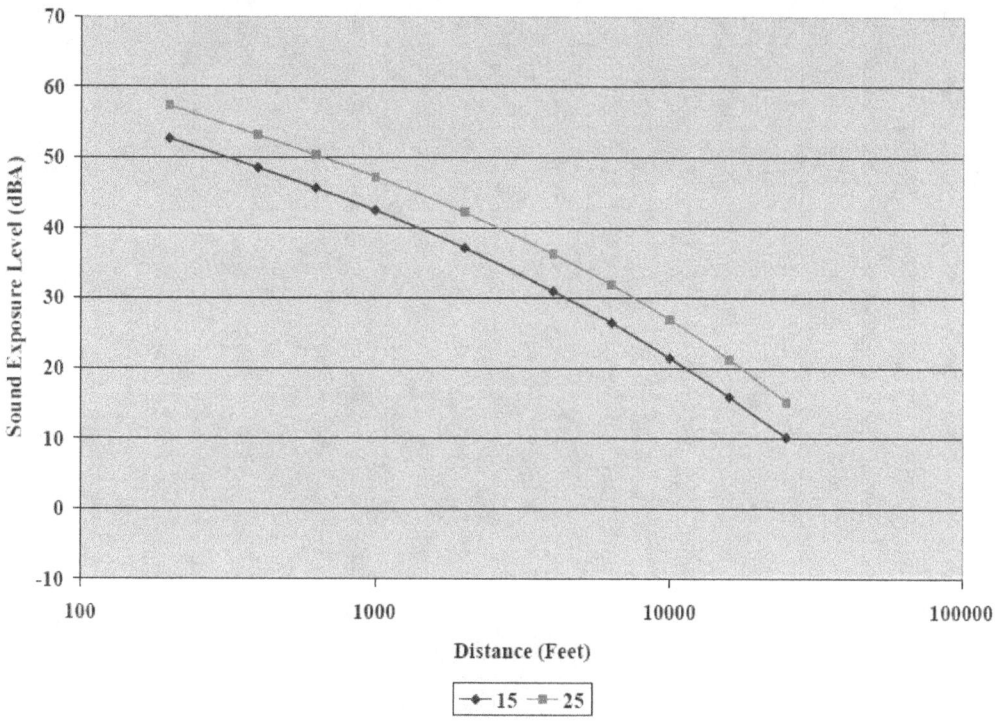

Figure 137. Yellowstone Expedition – Eleanor SEL Noise-Distance Curve

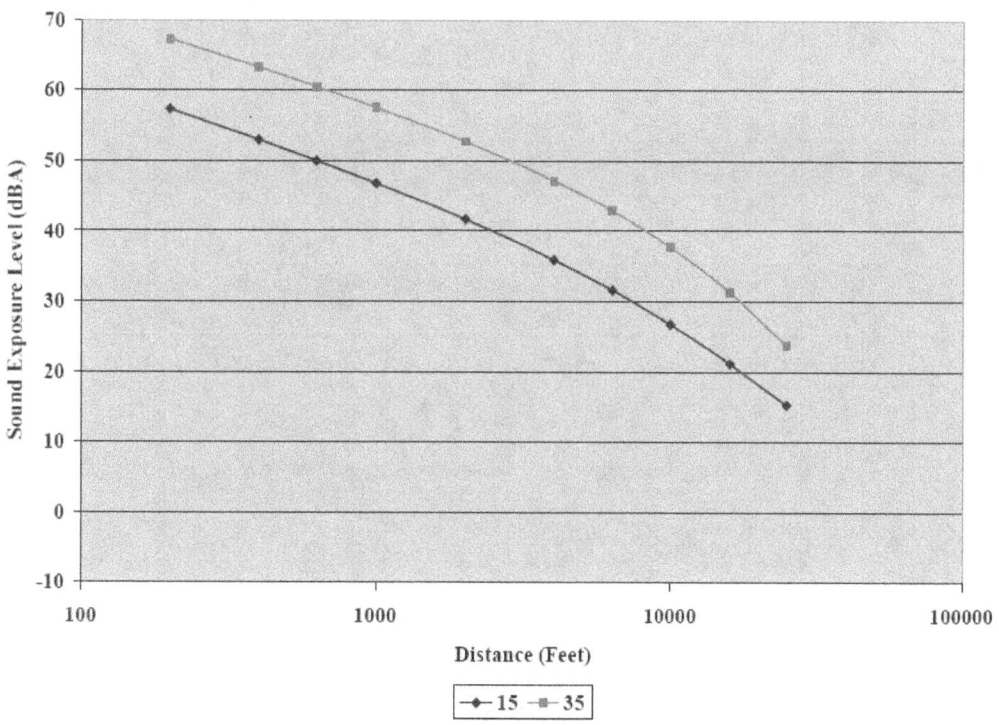

Figure 138. See Yellowstone Tours #6 SEL Noise-Distance Curve

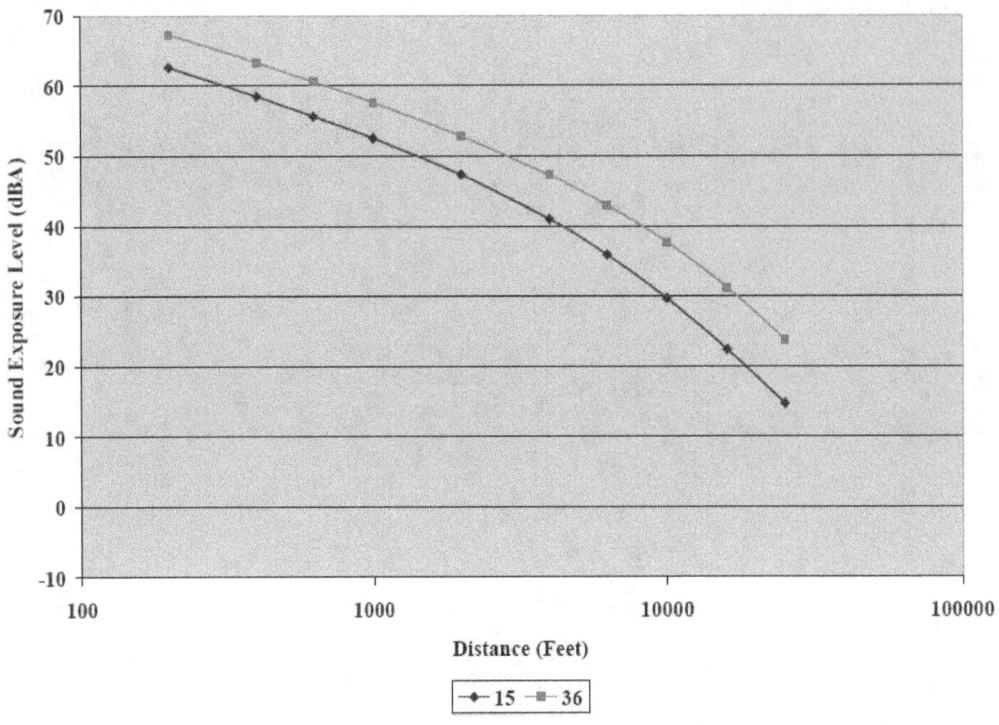

Figure 139. See Yellowstone Tours #9 SEL Noise-Distance Curve

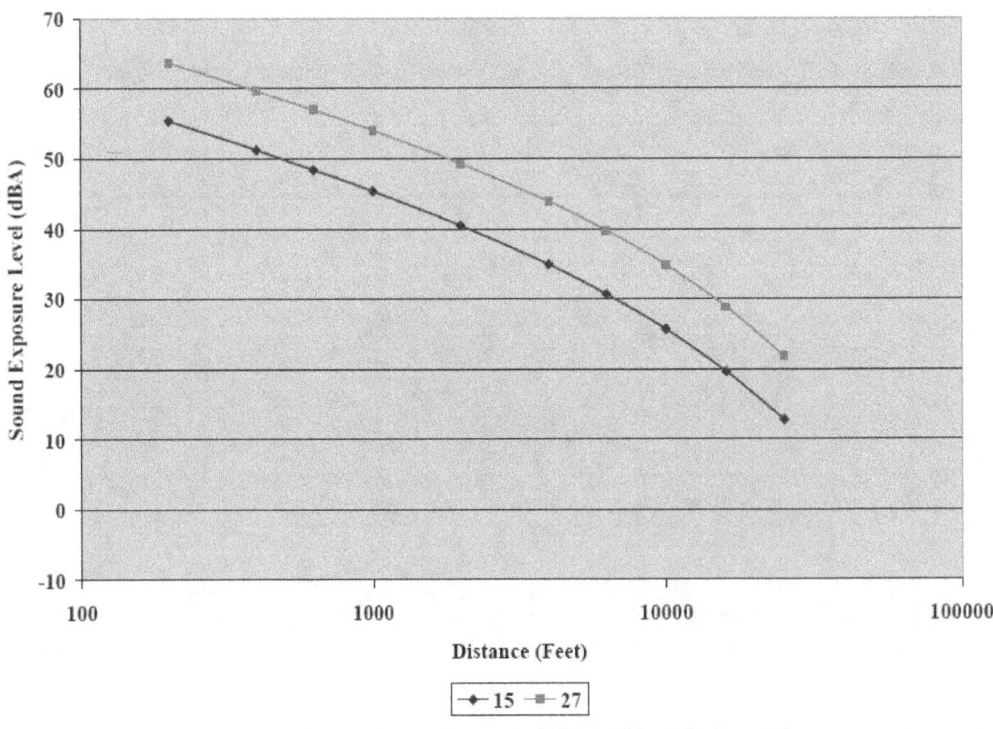

Figure 140. Buffalo Bus Touring #3 SEL Noise-Distance Curve

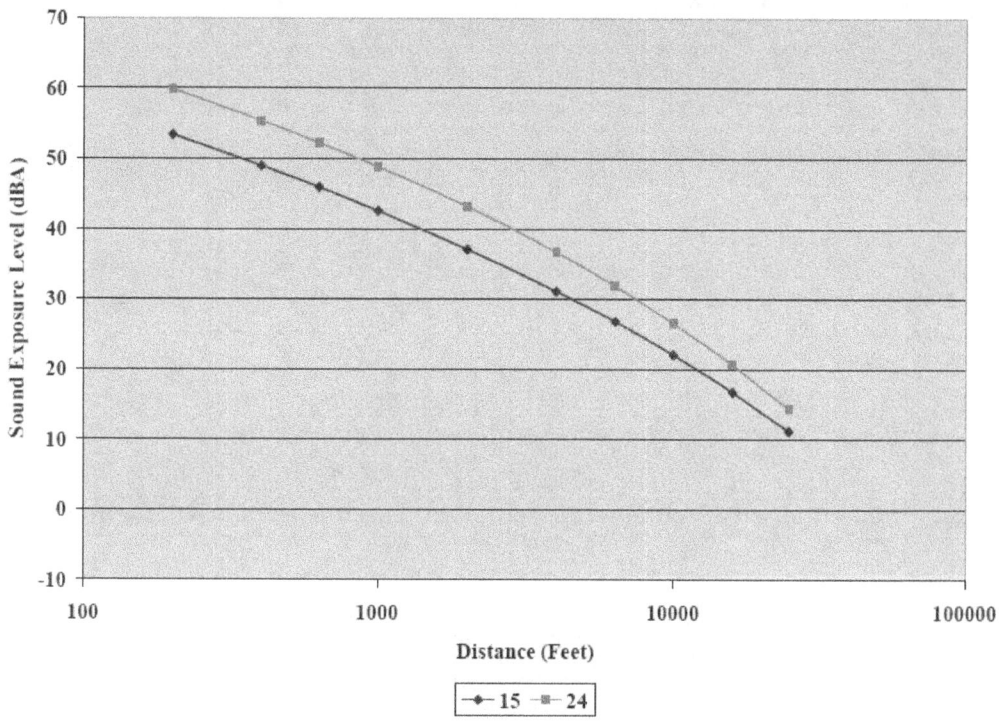

Figure 141. Buffalo Bus Touring #4 SEL Noise-Distance Curve

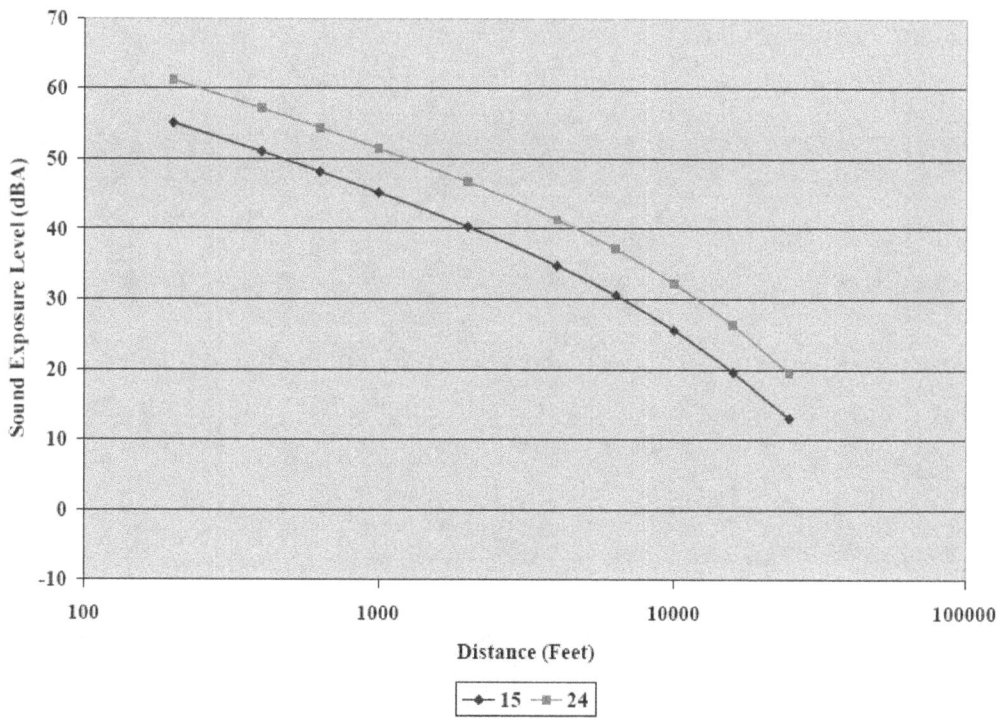

Figure 142. Buffalo Bus Touring T2 SEL Noise-Distance Curve

This page intentionally left blank

Appendix J: Spectral Data for Modeling

Spectral data for all events at 1,000 feet at the time of L_{ASmx} can be seen in Table 68 through Table 73. Figure 143 though Figure 158 show these data graphically.

Table 68. Spectral Numerical Values for All Events at 1,000 Feet at the Time of L_{ASmx} (part 1)

Vehicle		Xanterra 713			Yellowstone Expedition – Hayden		
Speed (mph)		15	25	32	15	20	30
		Spectra Sound Pressure Level at 1000 feet at time of LASmx (dB)					
Frequency (Hz)	12.5	25.8	32.2	25.1	31.1	34.3	27.6
	16	33.7	33.4	33.5	28.6	35.9	26.3
	20	38.6	38.9	42.2	32.2	34.3	29.3
	25	38.1	45.9	42.2	31.3	37.5	35.5
	31.5	43.9	37.0	48.7	29.9	35.3	33.9
	40	35.4	35.2	39.2	30.1	33.8	32.1
	50	37.6	44.4	37.7	34.1	36.4	38.2
	63	40.6	37.9	40.6	42.1	38.0	37.1
	80	45.7	42.5	36.3	37.4	42.3	36.2
	100	34.7	41.5	39.5	32.0	38.8	37.0
	125	35.5	45.3	37.3	36.8	33.0	42.2
	160	36.2	35.9	46.1	33.6	38.0	41.6
	200	39.9	44.5	38.0	32.9	37.1	38.9
	250	36.2	39.9	42.2	30.2	32.1	40.0
	315	30.6	34.7	39.0	28.1	28.8	40.3
	400	31.6	39.1	40.3	27.9	28.1	39.1
	500	28.9	38.9	44.3	27.6	28.5	37.6
	630	29.1	42.9	40.5	24.1	26.1	34.8
	800	29.2	38.6	40.8	22.7	25.6	32.1
	1000	27.5	37.0	42.4	21.9	25.3	31.0
	1250	25.6	36.8	39.0	19.9	22.6	29.3
	1600	24.4	35.4	38.5	18.1	21.3	25.9
	2000	21.1	31.6	35.2	16.5	20.3	24.9
	2500	18.5	28.4	32.8	14.1	17.5	22.6
	3150	14.9	24.8	30.1	11.3	14.3	20.0
	4000	10.0	18.8	25.8	6.5	9.1	15.3
	5000	9.9	16.1	23.9	4.6	6.4	14.0
	6300	3.3	9.0	17.4	-2.8	-0.2	8.0
	8000	-6.0	-1.0	10.4	-12.4	-10.3	0.6
	10000	-15.1	-11.7	0.8	-23.4	-22.2	-8.6

Table 69. Spectral Numerical Values for All Events at 1,000 Feet at the Time of L_{ASmx} (part 2)

Vehicle		Yellowstone Snowcoach – SNOVAN5			Xanterra 430		Xanterra 537	
Speed (mph)		15	25	35	15	23	10	16
		Spectra Sound Pressure Level at 1000 feet at time of LASmx (dB)						
Frequency (Hz)	12.5	30.5	30.5	38.3	34.4	35.1	33.5	34.2
	16	41.2	36.7	42.0	37.3	36.8	36.1	39.2
	20	36.5	34.7	44.9	35.5	37.9	41.8	42.8
	25	33.0	29.6	42.9	37.3	36.4	37.5	40.3
	31.5	32.7	30.4	40.0	36.2	35.2	37.1	41.8
	40	30.3	32.6	41.0	32.0	35.7	38.8	40.2
	50	31.4	33.9	40.4	34.8	35.4	47.6	39.0
	63	32.8	33.4	39.0	35.7	33.4	37.0	41.4
	80	41.6	30.2	36.2	43.7	32.3	42.9	44.6
	100	28.8	35.4	34.6	38.0	37.3	40.3	43.4
	125	27.0	42.0	34.6	29.6	42.3	45.4	40.1
	160	33.2	36.6	42.1	37.7	33.0	37.6	40.5
	200	30.4	34.0	46.7	33.4	36.1	42.4	39.8
	250	33.6	37.6	43.1	35.8	44.0	39.2	43.0
	315	34.1	37.8	47.1	37.2	40.5	40.2	44.9
	400	31.0	34.8	48.5	35.6	39.8	37.7	39.9
	500	27.4	32.3	44.4	32.8	36.8	34.5	39.6
	630	27.0	30.6	41.3	28.8	34.9	32.7	37.8
	800	26.4	30.3	40.5	27.2	33.9	31.7	35.0
	1000	25.3	29.4	38.8	22.5	30.5	29.7	37.1
	1250	22.3	28.1	35.9	19.6	28.9	28.0	37.7
	1600	21.0	26.2	34.3	17.9	26.1	31.5	37.7
	2000	20.0	25.2	32.8	16.3	25.2	30.9	37.0
	2500	18.1	22.1	31.3	14.4	24.1	28.2	34.0
	3150	16.3	18.9	29.0	12.2	20.7	24.7	30.9
	4000	12.5	15.5	24.3	9.5	16.9	23.1	28.2
	5000	11.5	14.1	22.7	6.2	14.4	20.8	27.0
	6300	4.8	9.1	16.3	0.1	8.0	6.9	17.5
	8000	-3.0	2.8	8.6	-6.8	0.3	-0.9	6.2
	10000	-11.4	-6.2	0.1	-15.8	-8.7	-14.8	-6.7

Table 70. Spectral Numerical Values for All Events at 1,000 Feet at the Time of L_{ASmx} (part 3)

Vehicle		Xanterra 707		Xanterra 709		Xanterra 710	
Speed (mph)		15	29	15	28	15	32
		Spectra Sound Pressure Level at 1000 feet at time of LASmx (dB)					
Frequency (Hz)	12.5	46.1	35.3	26.8	27.2	24.0	26.6
	16	48.9	39.8	31.2	31.7	30.3	36.3
	20	48.3	45.5	39.3	34.7	34.6	38.0
	25	46.8	47.5	42.5	49.4	37.0	41.4
	31.5	45.5	52.7	39.4	49.6	38.8	47.3
	40	44.6	44.6	38.1	40.8	38.4	39.5
	50	43.3	45.7	37.8	43.7	36.7	45.1
	63	43.4	44.7	40.7	43.8	38.5	42.3
	80	45.5	42.5	42.2	46.0	39.1	41.9
	100	40.0	42.3	34.5	39.8	36.2	40.5
	125	36.4	41.7	37.3	46.1	35.6	41.2
	160	37.9	43.4	36.4	44.4	34.7	44.1
	200	39.8	45.4	33.0	41.4	37.5	41.7
	250	35.9	45.0	31.0	45.0	33.7	45.9
	315	32.6	40.6	30.2	45.0	31.8	40.8
	400	31.9	43.2	28.3	44.4	29.9	41.1
	500	30.9	45.7	27.4	43.5	28.2	38.8
	630	30.7	48.2	28.6	40.4	27.4	40.6
	800	31.3	43.9	29.4	39.6	27.2	40.3
	1000	30.0	42.0	27.5	38.1	26.7	40.1
	1250	28.8	39.7	27.0	41.4	23.3	37.6
	1600	28.0	39.3	26.2	39.4	22.8	36.4
	2000	26.1	37.1	24.9	37.4	20.7	34.1
	2500	23.2	35.0	21.8	36.2	19.4	31.5
	3150	19.5	31.7	21.7	34.8	16.5	28.2
	4000	15.2	27.7	23.1	33.2	12.9	24.5
	5000	15.0	25.6	24.0	33.2	13.8	22.5
	6300	8.6	19.7	17.4	27.8	7.9	16.3
	8000	1.0	12.2	10.2	19.1	-2.7	9.7
	10000	-7.3	3.3	-5.8	5.4	-9.3	0.1

Table 71. Spectral Numerical Values for All Events at 1,000 Feet at the Time of L_{ASmx} (part 4)

Vehicle		Alpen Guide – Kitty		Yellowstone Snowcoach – SNOVAN4		Yellowstone Expedition – Eleanor	
Speed (mph)		15	31	15	33	15	25
		Spectra Sound Pressure Level at 1000 feet at time of LASmx (dB)					
Frequency (Hz)	12.5	30.4	29.4	40.2	43.0	33.5	26.1
	16	31.6	32.3	44.2	44.8	30.7	30.4
	20	35.7	39.1	41.0	44.4	29.8	37.4
	25	37.3	39.3	37.6	43.0	34.6	34.5
	31.5	35.9	38.2	36.6	40.4	28.8	32.9
	40	34.8	37.0	31.8	39.1	31.2	37.2
	50	34.3	37.1	34.1	42.1	33.2	36.1
	63	41.6	37.5	34.4	39.9	39.7	38.0
	80	42.4	34.7	42.1	34.6	39.9	33.8
	100	39.7	35.0	30.8	32.2	32.2	35.2
	125	31.7	36.8	25.9	33.6	34.3	41.9
	160	29.5	40.9	32.6	41.9	33.5	36.4
	200	28.0	32.9	29.8	42.9	32.5	37.5
	250	26.7	33.1	33.3	40.3	31.9	37.6
	315	24.5	33.2	35.0	48.6	31.7	36.1
	400	22.0	29.2	31.7	45.4	32.0	36.0
	500	22.2	29.9	28.3	43.2	29.4	34.6
	630	22.1	30.6	28.0	40.3	27.3	32.5
	800	20.9	29.3	27.1	39.9	26.0	30.8
	1000	21.9	28.6	23.9	36.7	27.6	31.1
	1250	21.6	29.1	20.2	32.7	28.4	30.6
	1600	17.7	25.5	18.4	29.9	26.0	28.3
	2000	14.8	23.8	17.7	28.3	22.5	25.3
	2500	13.6	21.4	12.5	24.2	19.6	22.2
	3150	8.9	17.3	7.4	19.8	15.4	18.6
	4000	2.7	11.2	1.2	13.8	9.6	13.8
	5000	-0.4	7.2	-2.9	10.0	8.0	12.2
	6300	-6.0	0.2	-10.7	2.7	-0.2	5.6
	8000	-16.9	-9.1	-23.2	-9.7	-5.5	-2.2
	10000	-27.4	-20.2	-36.4	-22.5	-16.0	-11.3

Table 72. Spectral Numerical Values for All Events at 1,000 Feet at the Time of L_{ASmx} (part 5)

Vehicle		See Yellowstone Tours #6		See Yellowstone Tours #9		Buffalo Bus Touring #3	
Speed (mph)		15	35	15	36	15	27
Frequency (Hz)		Spectra Sound Pressure Level at 1000 feet at time of LASmx (dB)					
	12.5	28.0	34.4	34.3	35.5	25.5	23.7
	16	38.2	36.5	36.0	39.5	31.2	30.8
	20	48.5	41.1	41.2	41.8	39.6	36.9
	25	50.3	44.6	43.1	40.1	43.0	39.6
	31.5	49.1	42.4	44.9	41.4	43.4	40.3
	40	40.9	43.4	40.1	40.4	37.9	35.4
	50	40.0	40.1	33.7	38.8	36.2	42.0
	63	34.1	39.3	44.6	32.8	38.9	37.1
	80	47.4	33.0	31.2	34.9	35.8	29.8
	100	31.7	36.3	32.3	31.9	30.2	34.5
	125	29.7	37.6	34.9	35.3	30.6	41.9
	160	37.6	41.7	34.2	49.2	35.2	37.4
	200	34.2	46.1	39.0	44.8	32.2	36.5
	250	38.1	44.1	39.9	42.7	36.0	48.1
	315	39.0	48.9	41.3	50.0	38.4	44.8
	400	37.5	48.9	40.6	45.3	36.3	45.8
	500	34.5	45.0	39.3	48.8	33.0	42.2
	630	31.6	42.1	40.4	42.1	30.9	39.2
	800	31.3	39.9	44.8	40.9	29.4	37.3
	1000	30.4	37.7	39.4	38.8	26.5	34.7
	1250	29.8	35.4	35.3	34.6	24.7	31.1
	1600	27.5	33.2	32.5	31.7	23.5	28.3
	2000	25.3	30.7	29.9	30.1	22.4	26.4
	2500	23.7	28.8	28.1	28.2	20.3	25.4
	3150	24.3	27.1	25.7	25.8	18.5	23.6
	4000	19.8	23.1	18.0	21.5	14.3	19.3
	5000	18.7	20.8	15.9	19.5	12.5	18.0
	6300	12.7	15.5	8.8	13.1	5.9	11.6
	8000	7.2	8.2	1.8	6.9	-2.4	3.8
	10000	-0.8	-0.7	-7.4	-3.4	-10.4	-5.4

Table 73. Spectral Numerical Values for All Events at 1,000 Feet at the Time of L_{ASmx} (part 6)

Vehicle		Buffalo Bus Touring #4		Buffalo Bus Touring T2	
Speed (mph)		15	24	15	24
		Spectra Sound Pressure Level at 1000 feet at time of LASmx (dB)			
Frequency (Hz)	12.5	38.4	28.7	27.4	30.6
	16	41.1	30.6	35.8	36.7
	20	48.3	40.2	36.4	37.8
	25	44.8	42.5	45.8	40.7
	31.5	39.9	42.9	53.3	38.4
	40	44.0	42.9	40.1	38.1
	50	49.5	38.5	37.2	41.2
	63	32.4	43.1	39.9	32.0
	80	31.5	37.3	37.9	31.4
	100	32.3	31.9	29.9	37.7
	125	30.3	34.9	28.9	42.4
	160	31.6	35.2	36.0	33.6
	200	35.0	40.2	32.5	36.1
	250	33.3	37.6	36.2	45.9
	315	32.8	36.8	38.1	42.6
	400	30.2	34.1	37.0	43.4
	500	26.8	32.2	33.1	39.5
	630	27.0	33.1	30.1	36.1
	800	26.6	34.4	30.0	36.3
	1000	25.7	34.2	27.7	32.9
	1250	25.1	33.4	25.1	29.3
	1600	25.3	31.9	23.2	26.5
	2000	24.4	30.7	20.3	24.0
	2500	21.2	27.1	19.9	23.7
	3150	19.9	25.0	17.3	21.6
	4000	17.3	23.4	14.0	17.6
	5000	15.1	21.8	11.0	16.0
	6300	9.1	15.3	6.1	10.9
	8000	2.5	9.7	-1.8	3.2
	10000	-5.9	1.4	-10.4	-6.1

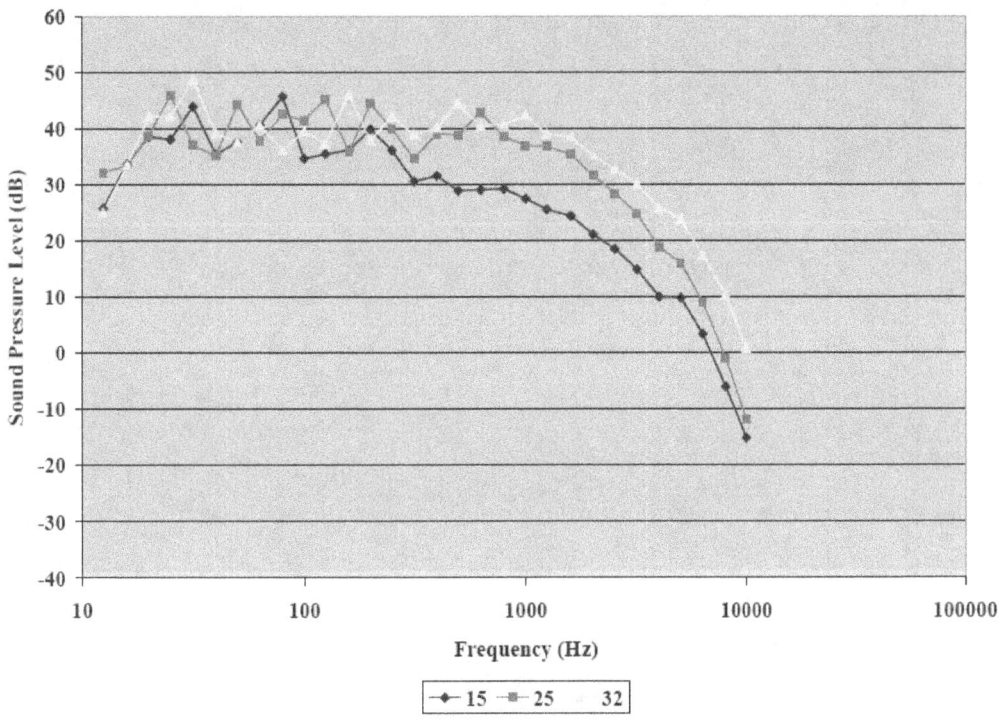

Figure 143. Xanterra 713 Spectra at 1,000 Feet at the Time of L_{ASmx}

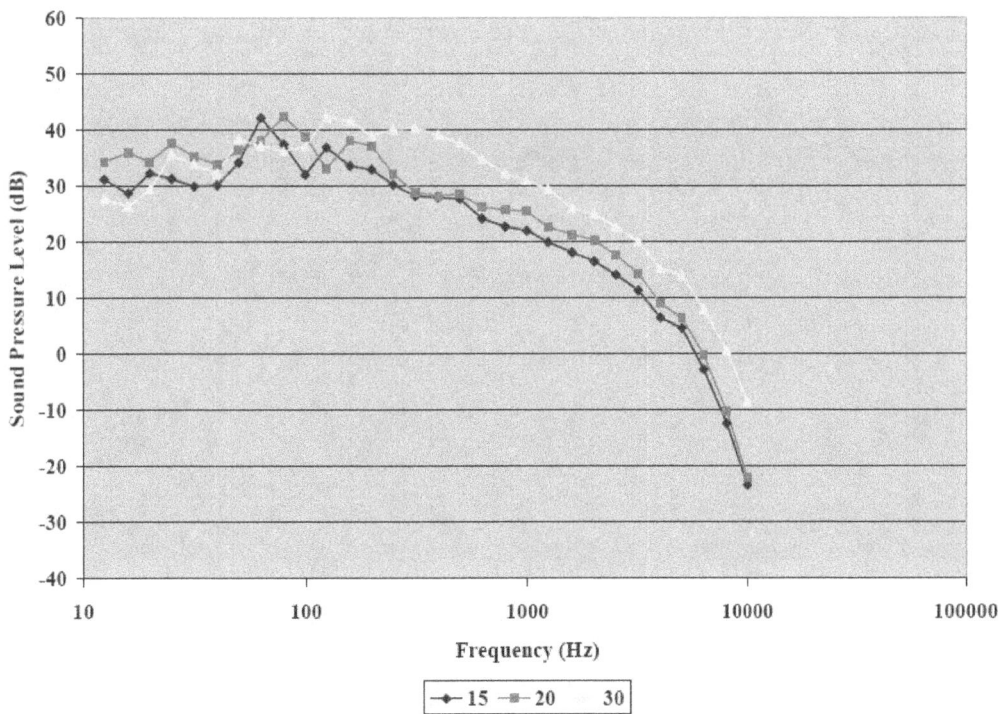

Figure 144. Yellowstone Expedition – Hayden Spectra at 1,000 Feet at the Time of L_{ASmx}

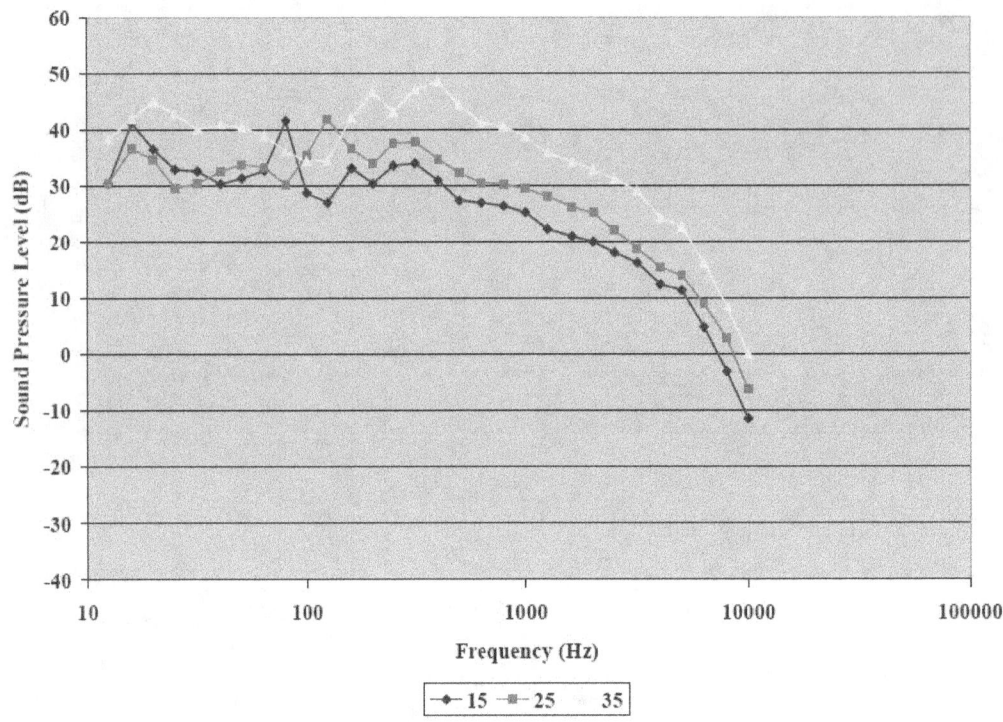

Figure 145. Yellowstone Snowcoach – SNOVAN5 Spectra at 1,000 Feet at the Time of L_{ASmx}

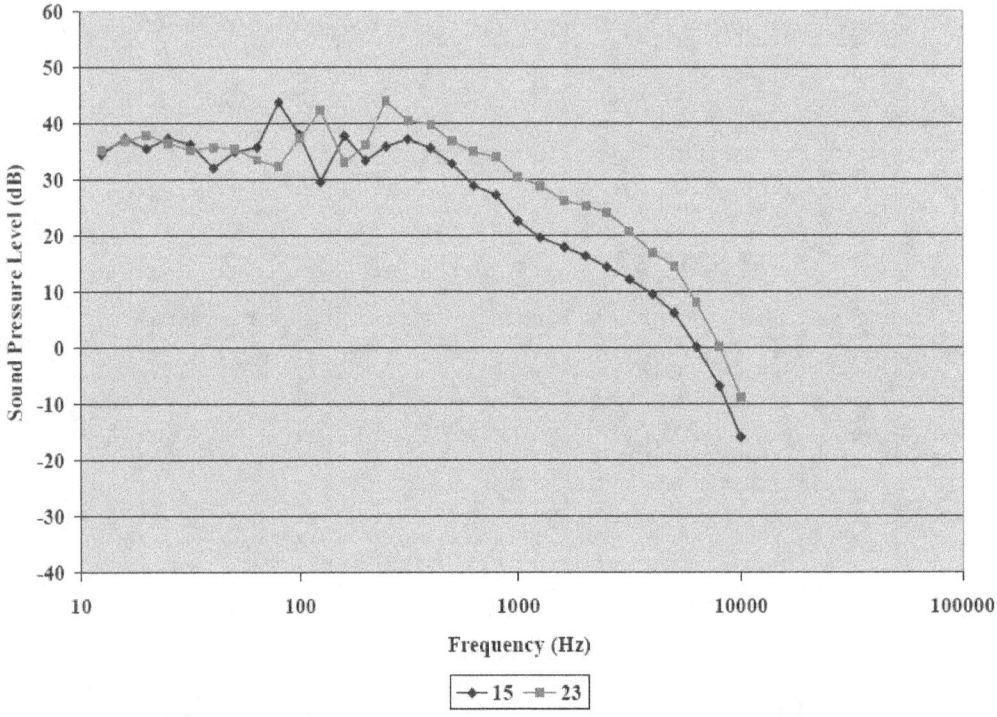

Figure 146. Xanterra 430 Spectra at 1,000 Feet at the Time of L_{ASmx}

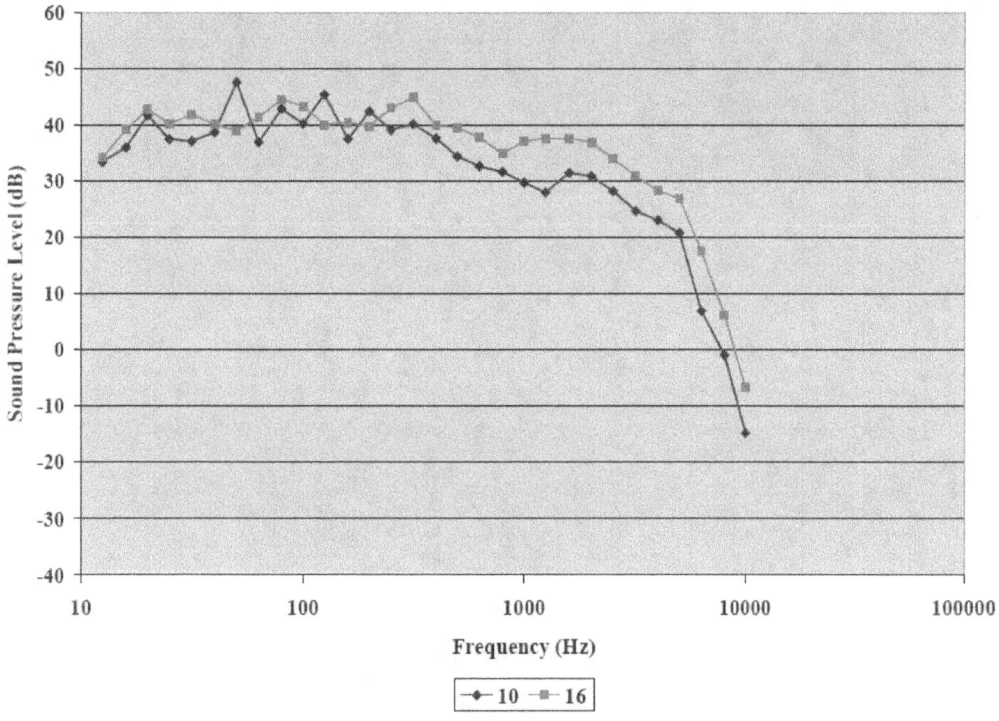

Figure 147. Xanterra 537 Spectra at 1,000 Feet at the Time of L_{ASmx}

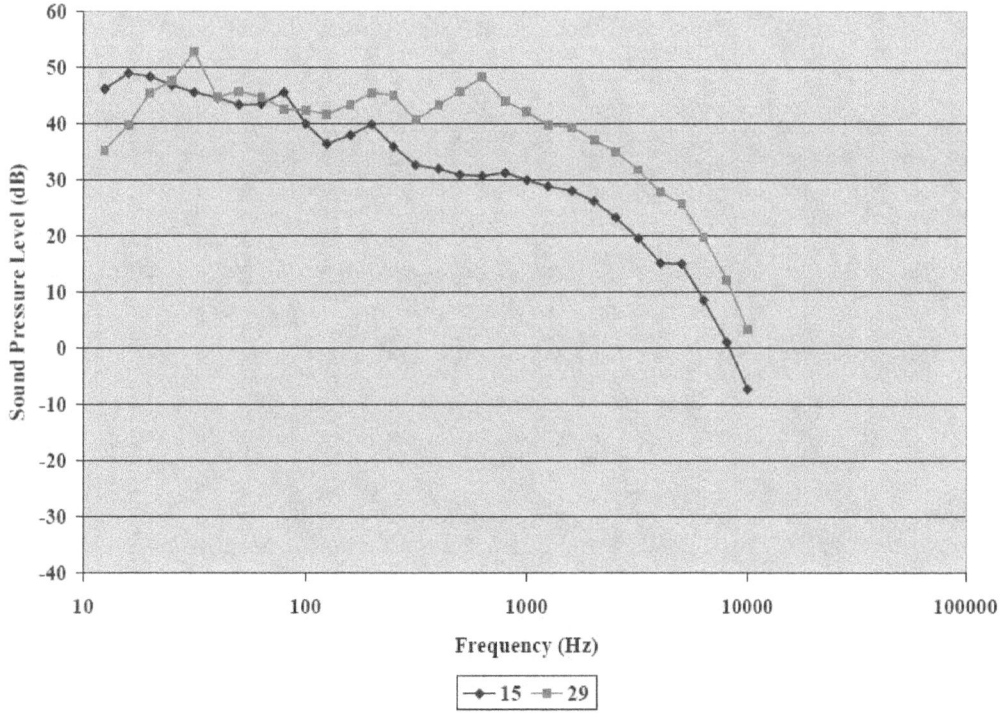

Figure 148. Xanterra 707 Spectra at 1,000 Feet at the Time of L_{ASmx}

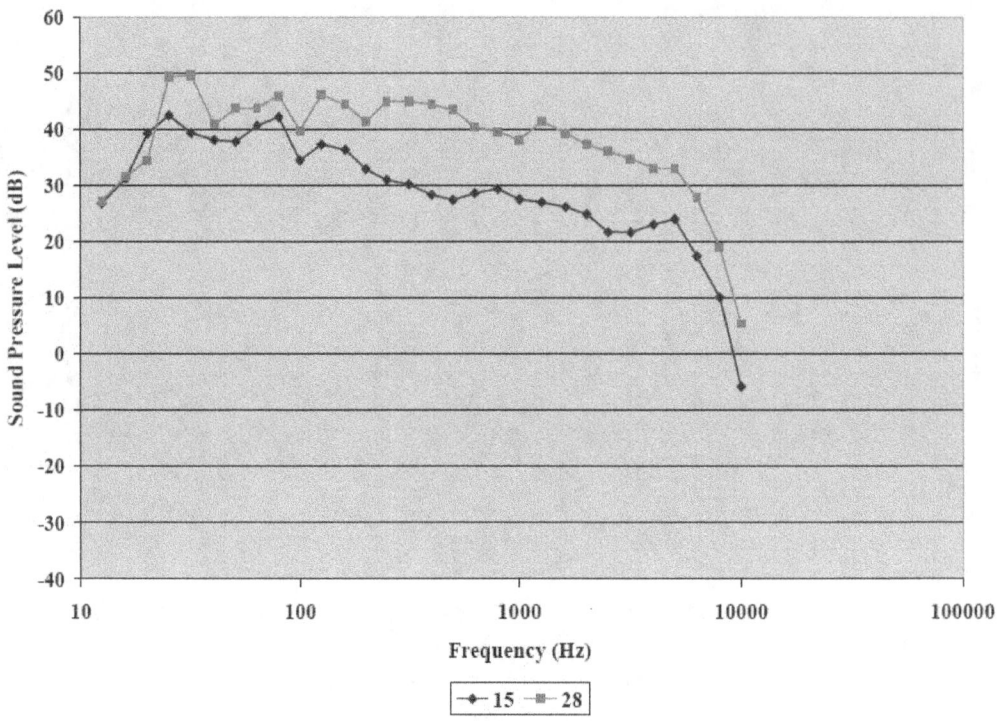

Figure 149. Xanterra 709 Spectra at 1,000 Feet at the Time of L_{ASmx}

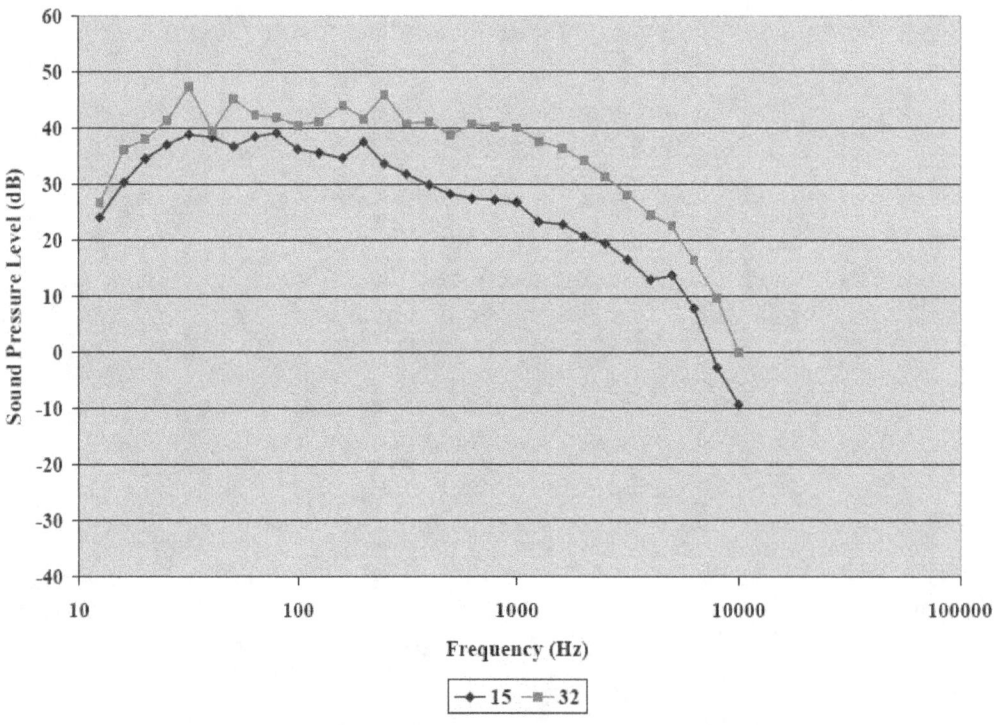

Figure 150. Xanterra 710 Spectra at 1,000 Feet at the Time of L_{ASmx}

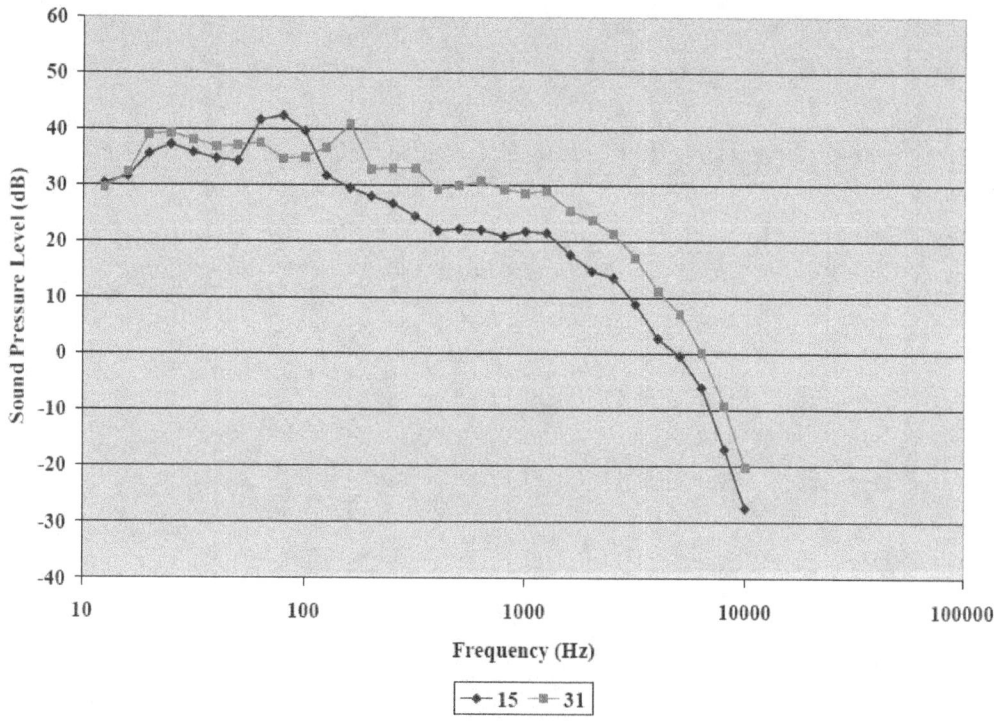

Figure 151. Alpen Guide – Kitty Spectra at 1,000 Feet at the Time of L_{ASmx}

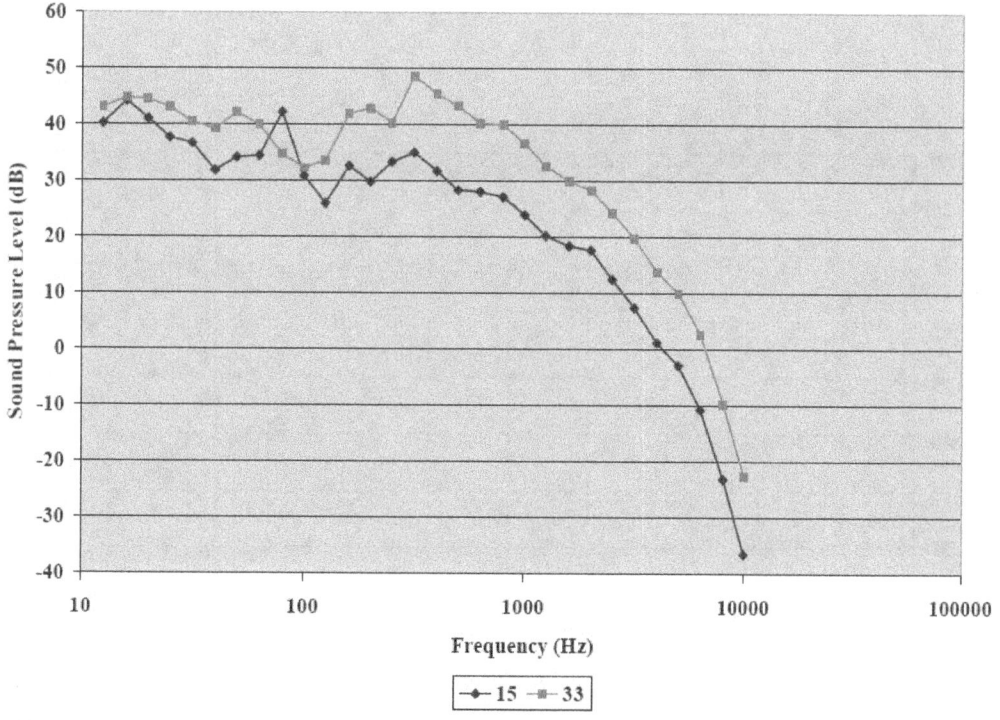

Figure 152. Yellowstone Snowcoach – SNOVAN4 Spectra at 1,000 Feet at the Time of L_{ASmx}

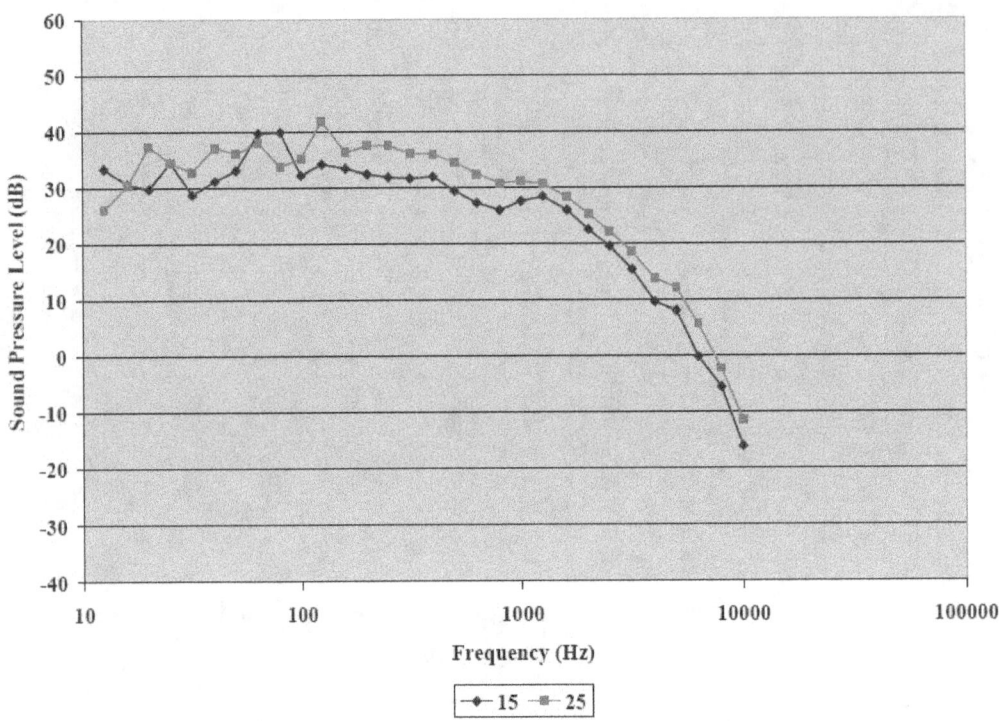

Figure 153. Yellowstone Expedition – Eleanor Spectra at 1,000 Feet at the Time of L_{ASmx}

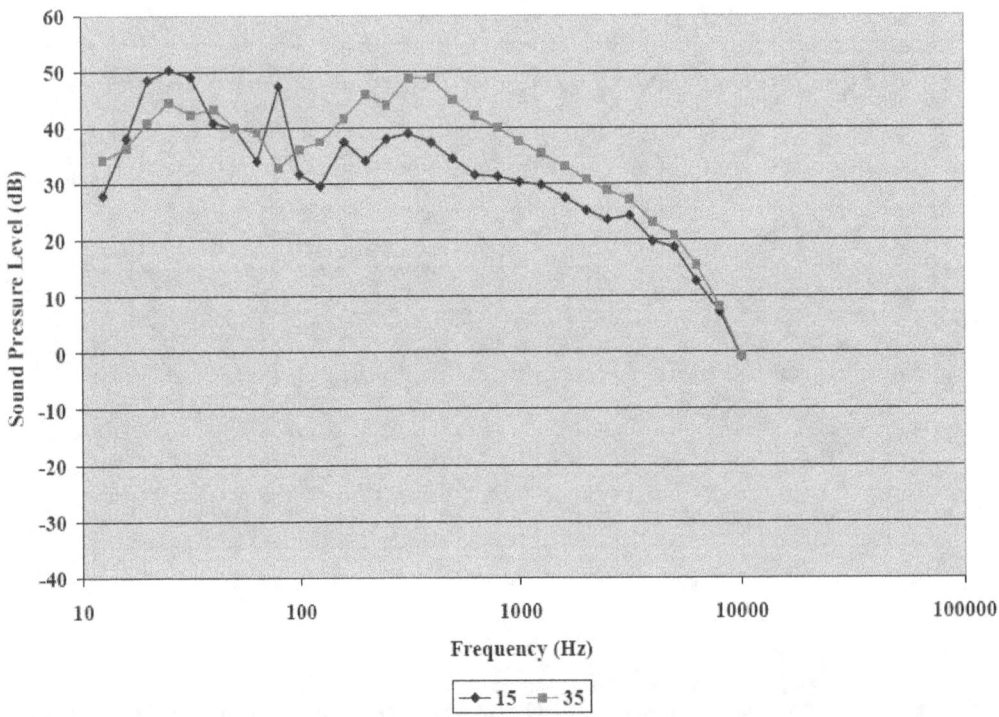

Figure 154. See Yellowstone Tours #6 Spectra at 1,000 Feet at the Time of L_{ASmx}

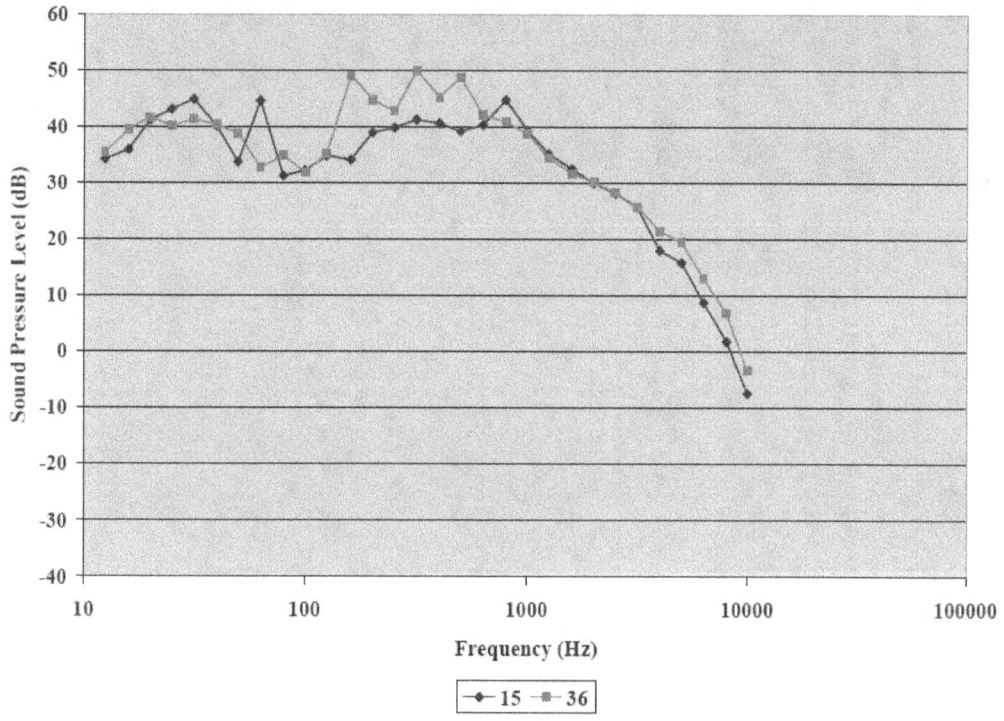

Figure 155. See Yellowstone Tours #9 Spectra at 1,000 Feet at the Time of L_{ASmx}

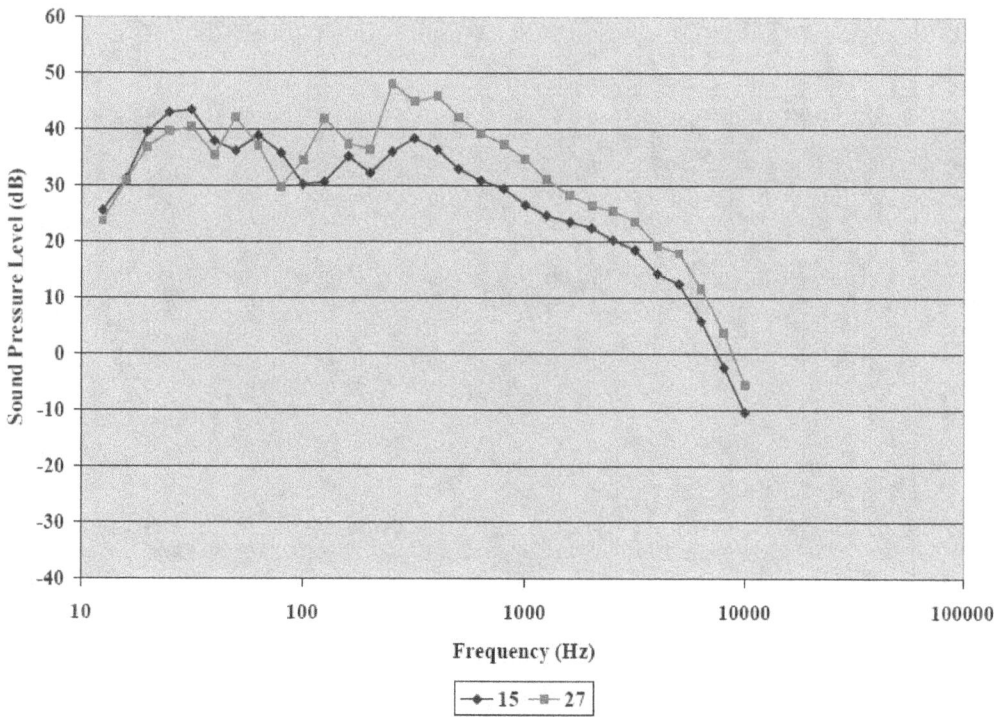

Figure 156. Buffalo Bus Touring #3 Spectra at 1,000 Feet at the Time of L_{ASmx}

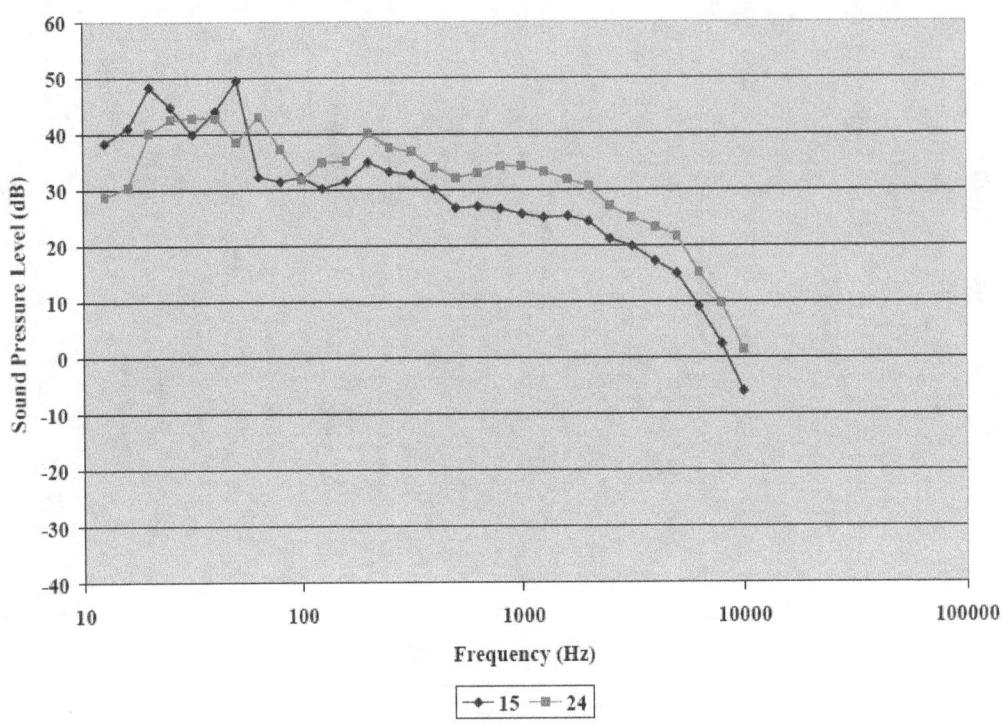

Figure 157. Buffalo Bus Touring #4 Spectra at 1,000 Feet at the Time of L_{ASmx}

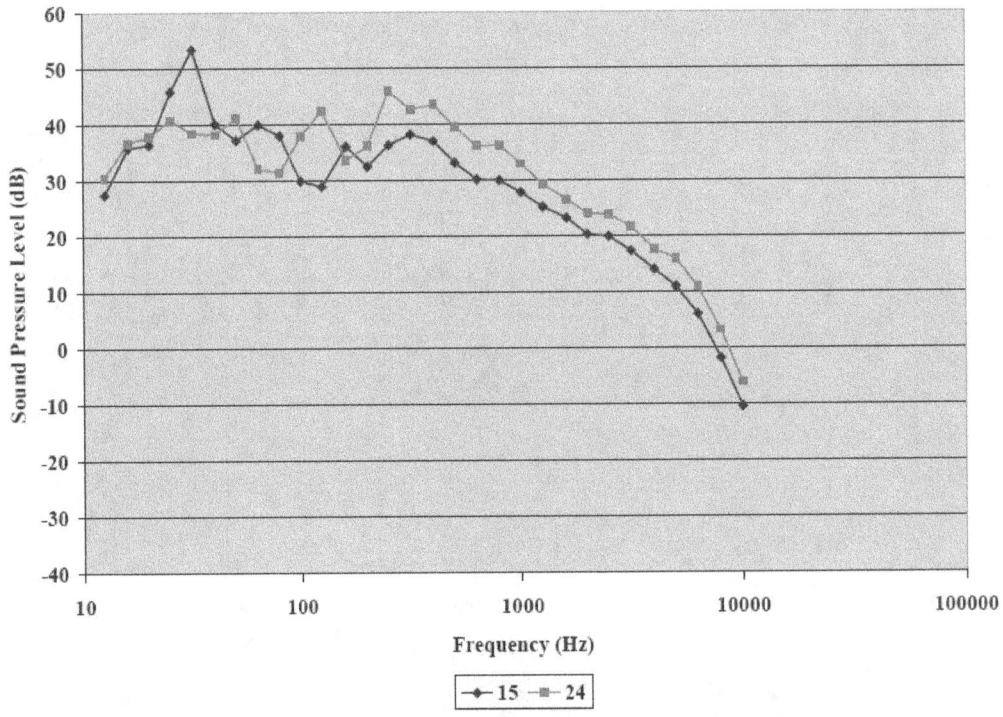

Figure 158. Buffalo Bus Touring T2 Spectra at 1,000 Feet at the Time of L_{ASmx}

References

1 Jason Ross and Christopher Menge, "Technical Report on Noise: Winter Use Plan Final Environmental Impact Statement," HMMH Report No. 295860.18, Harris Miller Miller & Hanson Inc. 15 New England Executive Park, Burlington, MA 01803, June 2001.

2 Christopher W. Menge and Jason C. Ross, "Draft Supplemental Report on Noise: Winter Use Plan Final Supplemental Environmental Impact Statement," HMMH Report No. 295860.400, Harris Miller Miller & Hanson Inc. 15 New England Executive Park, Burlington, MA 01803, October 2002.

3 National Park Service, "Preliminary Draft Alternatives – Winter Use Plans," Yellowstone and Grand Teton National Parks and John D. Rockefeller, Jr., Memorial Parkway, May 19th, 2006 Draft, provided to Volpe by NPS.

4 National Park Service, Winter Use Technical Documents, Yellowstone National Park, National Park Service, http://www.nps.gov/yell/technical/planning/winteruse/plan/, accessed on August 23, 2006.

5 Aaron L. Hastings, Gregg G. Fleming, Cynthia S.Y. Lee, "Modeling Sound Due To Over-Snow Vehicles In Yellowstone and Grand Teton National Parks," DOT-VNTSC-NPS-06-06, John A. Volpe National Transportation System Center, Acoustics Facility, Cambridge, MA 02142-1093, October 2006.

6 Gulding, Olmsted, Bryan, Mirsky, Fleming, D'Aprile, and Gerbi, "INM User's Guide," FAA-AEE-99-03, Federal Aviation Administration, Office of the Environment and Energy, John A. Volpe National Transportation System Center, Acoustics Facility, Cambridge, MA 02142-0193, September 1999

7 Jeffery R. Olmstead, Gregg G. Fleming, John M. Gulding, Christopher J. Roof, Paul J. Gerbi, and Amanda S. Rapoza, "INM Technical Manual," FAA-AEE-02-01, Federal Aviation Administration, Office of the Environment and Energy, John A. Vole National Transportation System Center, Acoustics Facility, Cambridge, MA 02142-0193, January 2002.

8 Christopher W. Menge, Christopher F. Rosano, Grant S. Anderson, Christopher J. Bajdek, "FHWA TRAFFIC NOISE MODEL (FHWA TNM®) TECHNICAL MANUAL" DOT-VNTSC-FHWA-98-2, John A. Volpe National Transportation System Center, Acoustics Facility, Cambridge, MA 02142-1093, February 1998.

9 Aaron L. Hastings, Cynthia Lee, Paul Gerbi, Gregg G. Fleming, Shan Burson, "Development of a tool for modeling snowmobile and snowcoach noise in Yellowstone and Grand Teton National Parks" John A. Volpe National Transportation System Center, Acoustics Facility, Cambridge, MA, 02142, Submitted to the Noise Control Engineering Journal on January 5th, 2010.

10 Gregg G. Fleming, Amanda S. Rapoza, Cynthia S.Y. Lee, "Development of National Reference Energy Mean Emission Levels for the FHWA Traffic Noise Model (FHWA TNM®), Version 1.0" DOT-VNTSC-FHWA-96-2, John A. Volpe National Transportation System Center, Acoustics Facility, Cambridge, MA 02142-1093, November 1995

11 International Organization for Standardization, Committee ISO/TC 43, Acoustics, Sub-Committee SC1, Noise, Acoustics – "Attenuation of Sound during Propagation Outdoors – Part 1: Calculation of Absorption of Sound by the Atmosphere," ISO 9613-1, Geneva, Switzerland: International Organization for Standardization, 1993.

12 Edward J. Rickley, Gregg G. Fleming, and Christopher J. Roof, "Simplified Procedure for Computing the Absorption of Sound by the Atmosphere," Noise Control Engineering Journal, 55(6), November – December 2007.

13 Aaron L. Hastings, Chris J. Scarpone, Gregg G. Fleming, Cynthia S. Lee, "Exterior Sound Level Measurements of Over-Snow Vehicles at Yellowstone National Park," DOT-VNTSC-NPS-08-03, John A. Volpe National Transportation Systems Center, Acoustics Facility, Cambridge, MA 02142-1093, September 2008.

14 "Operational Sound Level Measurement Procedure for Snow Vehicles," Society of Automotive Engineers, SAE Surface Vehicle Recommended Practice J1161, November 1976, revised March 1983, revised April 2004.

15 Christopher W. Menge and Jason C. Ross, "Draft Supplemental Report on Noise: Winter Use Plan Final Supplemental Environmental Impact Statement," HMMH Report No. 295860.400, Harris Miller Miller & Hanson Inc. 15 New England Executive Park, Burlington, MA 01803, October 2002.

16 Aaron L. Hastings, Gregg G. Fleming, Cynthia S. Y. Lee, "Modeling Sound due to Over-Snow Vehicles in Yellowstone and Grand Teton National Parks" DOT-VNTSC-NPS-06-06, John A. Volpe National Transportation System Center, Environmental Measurement and Modeling Division, RTV-4F Acoustics Facility, Cambridge, MA 02142-1093, October 2006.

17 Lawrence E. Kinsler, Austin R. Frey, Alan B. Coppens and James V. Sanders, Fundamentals of Acoustics, John Wiley and Sons, New York, 1990.

18 Society of Automotive Engineers, Committee A-21, Aircraft Noise, "Standard Values of Atmospheric Absorption as a Function of Temperature and Humidity," Aerospace Research Report No. 866A, Warrendale, PA: Society of Automotive Engineers, Inc., March 1975